T0245302

EMOTIONAL AI AND HUMAN-AI INTERACTIONS IN SOCIAL NETWORKING

EMOTIONAL AI AND HUMAN-AI INTERACTIONS IN SOCIAL NETWORKING

Edited by

MUSKAN GARG
University of Florida, USA

DEEPIKA KOUNDAL
University of Petroleum and Energy Studies, Dehradun, Uttrakhand, India

Academic Press is an imprint of Elsevier
125 London Wall, London EC2Y 5AS, United Kingdom
525 B Street, Suite 1650, San Diego, CA 92101, United States
50 Hampshire Street, 5th Floor, Cambridge, MA 02139, United States
The Boulevard, Langford Lane, Kidlington, Oxford OX5 1GB, United Kingdom

ISBN: 978-0-443-19096-4

For information on all Academic Press publications visit our website at https://www.elsevier.com/books-and-journals

Publisher: Nikki Levy
Acquisitions Editor: Megan McManus
Editorial Project Manager: Sam Young
Production Project Manager: Neena S. Maheen
Cover Designer: Christian Bilbow

Typeset by TNQ Technologies

Contents

Contributors

Subramaniam Abbirooban
Department of Psychology (SF), PSG College of Arts and Science, Coimbatore, Tamil Nadu, India

K. Abhilash
Amity University Rajasthan, Kant Kalwar, Rajasthan, India

K. Abilash
Department of Psychology (SF), PSG College of Arts and Science, Coimbatore, Tamil Nadu, India

Kriti Ahuja
Department of Allied Sciences, School of Health Science and Technology, University of Petroleum & Energy Studies, Dehradun, Uttrarakhand, India

Mohammed Hasan Ali Al-Abyadh
Department of Special Education, College of Education in Wadi Alddawasir, Prince Sattam bin Abdulaziz University, Saudi Arabia; College of Education, Thamar University, Dhamar, Yemen

Vamsi Kumar Attuluri
Department of Medical Lab Technology, University Institute of Applied Health Sciences, Chandigarh University, Mohali, Punjab, India

Duvvi Roopesh Chandra
Department of Computer Science Engineering, School of Engineering and Technology, CHRIST University, Kengeri Campus, Bangalore, Karnataka, India

Mangi Lal Choudhary
B.N. College of Pharmacy, B. N. University, Udaipur, Rajasthan, India

Nabanita Choudhury
Faculty of Computer Technology, Assam Down Town University, Guwahati, Assam, India

Rumi Iqbal Doewes
Faculty of Sport, Universitas Sebelas Maret, Surakarta, Central Java, Indonesia

Divya Dwivedi
Woxsen University, Decision Sciences & Artificial Intelligence, Hyderabad, Telangana, India; Supreme Court of India, New Delhi, India

Syam Machinathu Parambil Gangadharan
Liverpool John Moores University, Liverpool, United Kingdom

Vivek Kumar Garg
Department of Medical Lab Technology, University Institute of Applied Health Sciences, Chandigarh University, Mohali, Punjab, India

Neelam Goel
Department of Information Technology, University Institute of Engineering & Technology, Panjab University, Chandigarh, Punjab, India

Umesh Gupta
SR University, Warangal, Telangana, India

Vinh Truong Hoang
Faculty of Computer Science, Ho Chi Minh City Open University, Ho Chi Minh City, Vietnam

H. Indu
Department of Education, Avinashilingam Institute for Home Science and Higher Education for Women, Coimbatore, Tamil Nadu, India

Vasu Jain
SCSET, Bennett University, The Times of India Group, Greater Noida, Uttar Pradesh, India

Paridhi Jain
Department of Psychology, Manipal University Jaipur, Dahmi Kalan, Rajasthan, India

Ruchi Joshi
Department of Psychology, Manipal University Jaipur, Dahmi Kalan, Rajasthan, India

Swathikiran K.K.
Department of Computer Science Engineering, School of Engineering and Technology, CHRIST University, Kengeri Campus, Bangalore, Karnataka, India

M. Keerthika
Department of Psychology (SF), PSG College of Arts and Science, Coimbatore, Tamil Nadu, India

Markus Krebsz
University of Stirling, Management School, Stirling, United Kingdom; Woxsen University, Decision Sciences & Artificial Intelligence, Hyderabad, Telangana, India

Kukatlapalli Pradeep Kumar
Department of Computer Science Engineering, School of Engineering and Technology, CHRIST University, Kengeri Campus, Bangalore, Karnataka, India

Mohit Kumar
Department of Computer Science & Engineering, Jaypee Institute of Information Technology, Noida, Uttar Pradesh, India

Sunil Kumar
Department of Medical Lab Technology, University Institute of Applied Health Sciences, Chandigarh University, Mohali, Punjab, India

M.K. Kuralamudhu
Department of Psychology (SF), PSG College of Arts and Science, Coimbatore, Tamil Nadu, India

Nughthoh Arfawi Kurdhi
Faculty of Mathematics and Natural Science, Sebelas Maret University, Surakarta, Central Java, Indonesia

G. Maheswari
Department of Education, Avinashilingam Institute for Home Science and Higher Education for Women, Coimbatore, Tamil Nadu, India

Islahuzzaman Nuryadin
Faculty of Sport, Universitas Sebelas Maret, Surakarta, Central Java, Indonesia

Rajesh Prasad
School of Computing, MIT Art, Design and Technology University, Pune, Maharashtra, India

Jayashree Prasad
School of Computing, MIT Art, Design and Technology University, Pune, Maharashtra, India

Sapta Kunta Purnama
Faculty of Sport, Universitas Sebelas Maret, Surakarta, Central Java, Indonesia

Nihar Ranjan
Information Technology, JSPM's Rajarshi Shahu College of Engineering, Pune, Maharashtra, India

Manvendra Singh
Department of Government and Public Administration, Lovely Professional University, Phagwara, Punjab, India

Suyesha Singh
Department of Psychology, Manipal University Jaipur, Dahmi Kalan, Rajasthan, India

Divya Singh
Bennett University, Greater Noida, Uttar Pradesh, India

Michael Moses Thiruthuvanathan
Department of Computer Science Engineering, School of Engineering and Technology, CHRIST University, Kengeri Campus, Bangalore, Karnataka, India

Gargi Trivedi
Bennett University, Greater Noida, Uttar Pradesh, India

M. Sundararaj Vasanth
Department of Psychology (SF), PSG College of Arts and Science, Coimbatore, Tamil Nadu, India

CHAPTER ONE

Introduction to social neuroscience

Sunil Kumar[1], Vivek Kumar Garg[1], Vamsi Kumar Attuluri[1] and Neelam Goel[2]

[1]Department of Medical Lab Technology, University Institute of Applied Health Sciences, Chandigarh University, Mohali, Punjab, India

[2]Department of Information Technology, University Institute of Engineering & Technology, Panjab University, Chandigarh, Punjab, India

1. Introduction

We can understand this with an example. Suppose one way is toward mental peace and other way toward the big awards, name, and fame, so which pathway is the best for us, we do not know until unless we have not experienced such kind of experience. According to research, a variety of social factors, such as life experiences, poverty, unemployment, and loneliness, might affect biomarkers connected to health (Kennedy et al., 2021; Michaels et al., 2022; Steptoe et al., 2004). So, humanity is playing a very important role here. So, it is very important to explore the situation in the real field. In order to better understand the mechanisms behind human thought and behavior, the emerging area of social neuroscience emphasizes the complimentary link between many levels of organization (such as molecular, cellular, system, individual, relational, collective, and societal). While social and behavioral conceptions and data are utilized to develop ideas of brain organization and function, biological concepts and theories of social behavior are informed and improved through various methods (Shute, 2009; Why loneliness is hazardous to your health - PubMed, n.d.). Social neuroscience is a branch of psychology. It's exciting to be a part of a developing concept, business, or academic topic, but it can also be stressful. Such concepts often foretell future possibilities and are so novel that most people have not ever heard of them. There is always a chance that the concept may fail (Eisenberger & Cole, 2012; Guo & Zhu, 2022; Shute, 2009). However, when sales rise, new goods are introduced, and new discoveries are made, the initial tension transforms into a frenzied thrill that goes along with any novel, ground-breaking hypotheses, proposals, or

Emotional AI and Human-AI Interactions in Social Networking
ISBN: 978-0-443-19096-4
https://doi.org/10.1016/B978-0-443-19096-4.00010-9

notions, regardless of the industry or profession. Jason Mitchell, a social neuroscientist at Harvard University, recalled When he initially began, "there was all kinds of angst about whether what we were doing was ever going to establish any kind of toehold in the profession, or whether it was some bizarre, freakish carnival" he writes in the article "Peering Inside the Social Brain." However, those days are long gone now, thanks to the quick development of brain scanning technology in both public and private laboratories. In other words, social neuroscience, one of the most cutting-edge fields of research, is heading in the right direction (Cacioppo et al., 2000; Chiao et al., 2022; Glaser & Kiecolt-Glaser, 1994; Zachariae, 2009). The quarterly journal "Social Neuroscience Bulletin," which was issued between the years 1988 and 1994, is where the phrase "social neuroscience" first appeared. Although neuroscience is a relatively young and unexplored topic of study for educators, it may provide crucial insights for the creation of innovative teaching strategies and practices, as well as, in some circumstances, validate good instructional practice already working in Neuroscience (McIntyre, 2015, pp. 53–68).

2. Applications or focus on the whole

Neuropsychology and neuroscience have mostly been concerned with studying human functioning on an individual basis up until recently. However, the majority of experts in these subjects are aware that human behavior does not happen in emptiness. Conducts take place in framework intricate such community arrangements, which have a significant impact on and frequently dictate the behaviors or responses of individuals worldwide and across all cultural contexts. The study of the brain networks that underpin emotions and motives has also grown exponentially during the past 10 years. Social neuroscience researchers want to identify the neurological mechanisms influencing both emotions and motives because they are aware that emotions influenced by social variables frequently lead individuals to act or perform in particular traditions. Besides this sort of thrill is actually happening at this moment, now among the most recent scientific disciplines is known as social neuroscience (Ansari et al., 2021). Social psychology, cognitive psychology, neuropsychology, and neuroscience are all integrated in the social neuroscience approach. Social neuroscience aims to uncover socio-emotional aspects of impact according to the interplay of the social, cognitive, and neurological parts of how humans work by combining the ideas and methodologies of these psychological disciplines. Stereotyping, bias, person perception, and social exclusion are some of the areas of social neuroscience currently being researched. Furthermore, several subjects that were formerly taboo for academics are now popular research themes in societal neuroscience, such as the reasons behind why individual's affection, hatred, or dislike one another. How does one feel for another person? Why are some individuals more conflicted than others? Why do certain people develop fervor for certain religions or ideologies? And how can deficiencies in the brain affect a person's capacity for social interaction? A leading "National Institute of Mental Health" (NIMH) researcher stated in the Knowledge editorial "Looking into the Communal Mind" that until a few years ago, discussions on the neurological underpinnings of affection, bond, and belief "simply essentially appeared absurd." According to Janine Simmons, director of the NIMH program for emotion, social behavior, and social cognition, neuroscientists have historically been primarily constrained by technological advancements. "These study systems must demonstrate beneficial, nonetheless it remained the accessible accessibility, preliminary near a period ago, of functional neuroimaging technologies that fueled an expansion in social neuroscience," the authors transcribe imaging revisions of single-cell footages of animal brains. Functional magnetic resonance imaging (fMRI) is a type of imaging technique that has been around since the 1980s, but it has just recently been more accessible to universities and other research labs due to equipment prices that have decreased. See neuroimaging for further details. The number of scholars interested in identifying the neural networks underlying complicated social interactions grew as more brain scanners like fMRI became available spread through the nation. With this gear readily available, research projects multiplied. "The Social and Affective Neuroscience Society" and the "Society for Social Neuroscience" are two scientific organizations that have emerged as a result of the developing discipline (Tsakalidis et al., 2022).

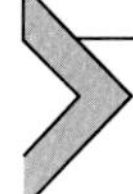

3. In contrast to explain more about the application, opportunities

For example, we may design online gaming to manage emotional responses, for example, people's intelligence can be engaged with the help of online gaming, virtual reality, online behaviour monitoring, (Marketing) hurdles, challenges, behaviour, and brain activity measuring by fMRI MEG, PET, EMG, TMS, GSR. In Fig. 1.1, we have demonstrated the

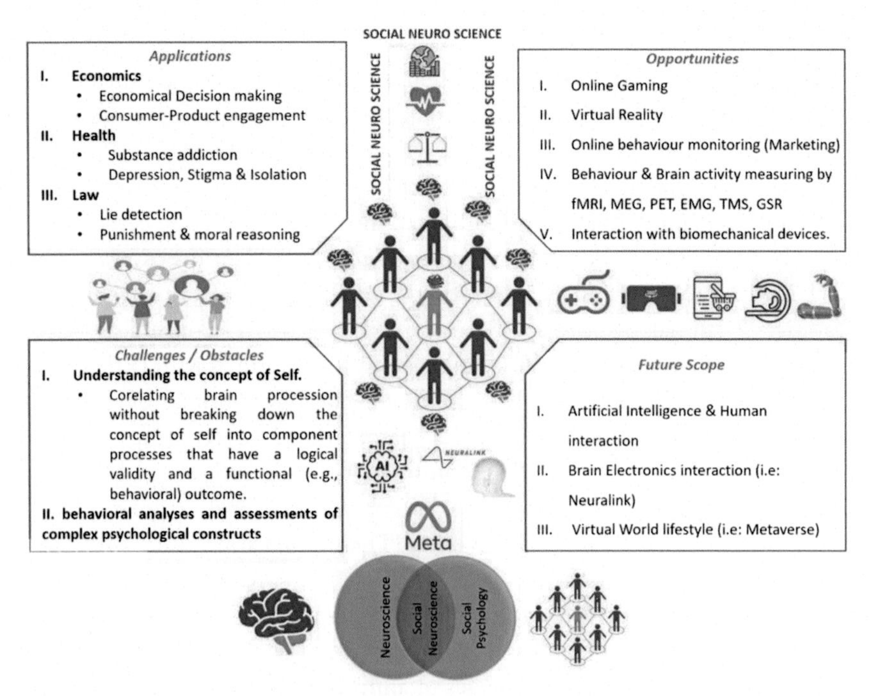

Figure 1.1 Different aspects of social neurosciences in relation to its applications, opportunities, challenges/obstacles, and future scope. Created with BioRender.com.

different aspects of social neurosciences in relation to its applications, opportunities, challenges/obstacles, and future scope.

These are the methods to detect and control emotions of the peoples and can be controlled up-to some extent.

4. Impact on society

An interdisciplinary discipline called social neuroscience investigates how social interactions and biological processes are related. Fundamentally, humans are a sociable creature as opposed to a lonely one. As a result, in addition to cities, civilizations, and cultures, humans also develop emergent social structures including couples, families, and groups. Many socioeconomic factors, such as life events, poverty, unemployment, and loneliness, have been shown in studies to affect biomarkers associated with one's health (Larkin et al., 2016). The phrase was made well-known by John Cacioppo and Gary Berntson in a 1992 study for the American Psychologist. Cacioppo and Berntson are typically regarded as the founding fathers of social neuroscience (Ansari & Coch, 2006). Affective neuroscience and cognitive neuroscience are closely related to social neuroscience, a relatively emerging subject that examines how the brain controls social relationships (Sapolsky, 2005; Zachariae, 2009). Social neuroscience, which is still in its infancy, is closely related to affective neuroscience and cognitive neuroscience and focuses on how the brain controls social relationships.

5. Application social neurosciences

Common neuroscience seeks to comprehend more fully the neurological mechanisms behind people's ideas, feelings, and behaviors in relation to their social environments. Academics, professionals, and experts from related professions have shown an interest in the topic of social neuroscience. Throughout the past several decades, the general public has come to rely more and more on social neuroscience ideas and techniques to explain, predict, and alter behavior. It is more crucial now than ever to consider the reliability of social neuroscience results and also the topics it can as well as cannot address (Tsakalidis et al., 2022). In this chapter, we look at how social neuroscience has impacted three fields with obvious societal relevance: health, economics, and law. In order to address concerns about extending neuroscientific findings to practical societal issues addressed in a prior study, we offer guidelines and best practices. The current book chapter emphasizes the value of researching health-related, economic, and legal decision-making in the social environment and illustrates how social neuroscience has participated in these three practical fields. While not complete, the research inquiries we provided as samples were picked because of their general theoretical underpinnings and approach to methodology. According to the findings presented in this article, different brain systems are activated when other people are engaged, and these mechanisms are influenced by the behavior, identity, and emotions elicited by the people with whom one interacts. Therefore, taking into account the social environment is essential for the paradigm's ecological validity (Kedia et al., 2017).

6. Where we can open the consultancy center?

Traditional neuroscience has long treated the nervous system as a standalone structure and generally disregarded the impacts of the social settings that many animal species, including humans, live in. In reality, social structures have a significant influence on how the body and brain function, as we now understand (Benton, 2010; Bruner, 1985). In order to develop theories of social processes and behavior in the social and behavioral sciences, social neuroscience applies concepts and methods from biology. It examines the biological mechanisms that underlie social processes and behavior, which is widely regarded as one of the major problem areas for the neurosciences in the 21st century (Coch & Ansari, 2009). The brain serves as the key governing organ as well as a flexible target for these factors, and these social factors have an ongoing interaction with neuronal, neuroendocrine, metabolic, and immunological processes on the brain and body (Goswami, 2009).

7. Possibilities

Although neuroscience is a relatively young and unexplored topic of study for educators, it may provide crucial insights for the creation of innovative teaching strategies and practices, as well as, in some circumstances, validate good classroom practices already in use. The study of how the brain works is called neuroscience. Research on brains after they had been removed from bodies limited early studies of the brain to structural investigations. With the brain acting as the primary regulatory organ and a flexible target for these factors, these social influences have an ongoing interaction with neuronal, neuroendocrine, metabolic, and immunological factors on the brain and body (Eisenberger & Cole, 2012). It was not until the 1990s that scientists were able to examine how brains performed by utilizing fMRI and positron emission tomography (PET) scans to map activity in the brain. Recent advances in technology have made it possible to study the neurological underpinnings of brain function using techniques like near-infrared spectroscopic imaging (NIRSI) and diffusion tensor imaging (DTI). As neuroscience continues to elucidate ideas that guide learning and, potentially, empower learners, it has aroused the curiosity of educators (Hammersley et al., 2021; Michaels et al., 2022).

8. Methods for investigating neural and social processes

Numerous techniques are employed in social neuroscience to look at how neurological and social processes interact (Murphy & Benton Stephen, 2010). Numerous techniques are employed in social neuroscience to look at how neurological and social processes interact such as fMRI, PET, transcranial magnetic stimulation (TMS), magnetoencephalography (MEG), facial electromyography (EMG), electroencephalography (EEG), event-related potentials (ERPs), electrocardiograms, and electromyography. Virtual reality (VR) and hormonal measurements have been added to these techniques in recent years. Animal models can also be used to examine the potential functions of particular brain regions, circuits, or functions (such as the reward system and drug addiction) (Ferrari et al., 2016). Moreover, quantitative meta-analyses are necessary to go beyond the constraints of individual studies, and neurodevelopmental research can advance our knowledge of the relationships between the brain and behavior (Hall et al., 2023). fMRI and EEG are the two techniques used in social neuroscience the most frequently. The imaging method used in fMRI has a cheap cost and a high spatial resolution. However, because of their poor temporal resolution, they are most useful for identifying the neural circuits that are employed in social studies. Since fMRI detects oxygenated blood levels, which pool in the brain regions that are active and require more oxygen (timing), it has a limited temporal resolution (Curley & Ochsner, 2017). Because of this, blood cannot be tested for precise time of activation during social tests because it takes longer for blood to reach the part of the brain that is active (Meyer et al., 2012). EEG is particularly helpful when an investigator is trying to map out a specific region of the brain that corresponds to a social construct they are researching. During these trials, it is quite challenging to separate these factors ("Neuroscience: Social Networks in the Brain," n.d.). Although EEGs have limited spatial resolution, they have great temporal resolution. Because the time of activation is quite accurate, yet identifying precise locations on the brain is challenging, researchers are trying to focus on certain places and regions, but they are also producing a lot of "noise." (Barrett & Satpute, 2013). Researchers have recently shown that TMS is the best technique for pinpointing the exact location of a process in the brain (Lewis et al., 2011). It is employed in social contexts. However, due to its high cost, this equipment is rarely utilized. The necessity of using correlations to understand research in social neuroscience has the potential to reduce the content validity of the findings. With the exception of TMS, the majority of these methods can only show relationships between social events and brain mapping (Ochsner et al., 2007). Self-reports are crucial in this regard. Also, this will lessen the likelihood of VooDoo linkages (correlations that are too high and over 0.8 which look like a correlation exists between two factors but actually is just an error in design and statistical measures). Using hormone-based testing that can infer causality is another way to avoid this con (Chiao et al., 2009). For instance, when patients receive placebos and oxytocin, we may assess how they interact with others on a social level. SCRs will assist in separating unconscious and conscious thoughts because they are the body's normal parasympathetic response to the outside world (Parkinson et al., 2017). Performance-based measures, like Implicit Association Test, that record response time and/or accuracy; observational measures, for example, preferential looking in infant related researches; psychological methods include self-report measures such as interviews and questionnaires (Spunt & Lieberman, 2014). Neurobiological methods are classified as measuring more external bodily responses, electrophysiological methods, hemodynamic measurements, and lesion methods. GSR (also known as skin conductance response (SCR)), eyeblink startle response, and facial EMG are examples of bodily response methods (Ward, 2012). EEG, ERPs, and single-cell recordings are all electrophysiological techniques. Instead of directly assessing brain activity, hemodynamic measurements like PET and fMRI measure changes in blood flow (Meyer & Lieberman, 2012).

9. Challenges and obstacles

Investigations of injured brains caused by natural events including strokes, severe traumas, tumors, neurosurgery, infection, or neurodegenerative illnesses have historically been conducted using lesion techniques. TMS may fall under this heading because of its capacity to produce a transient "virtual lesion." Specifically, TMS approaches stimulate a particular area of the brain while isolating it from the rest of the brain to simulate brain damage. This is especially useful for brain mapping, a critical method in social neuroscience for determining which brain areas are active during specific activities (Mitchell et al., 2005).

10. Future scope of social neurosciences

The study of human social interactions using game theoretical models will be the main emphasis of the area of social neuroscience, which will be linked to the even more recent field of neuroeconomics. We will talk about more current research on social emotions including love, compassion, retribution, and fairness (Berntson & Miller, 2018). The utilization of a multimethod and multidisciplinary research approach that integrates genetic, developmental, pharmacological, and computational components is also discussed, as well as future objectives for the study of interactive minds (van Rensburg & Adcock, 2016).

11. Role of social Networking

Because most people are not social by physical contact these days, social networking is playing a major role in social neurosciences. Because everyone is connected to each other through social websites and social apps, networking is playing a very important role in the future. fMRI and social network analysis demonstrate that the brain activates areas necessary for discerning mental states and intentions, as well as regions associated with spatial navigation and psychological distance, when monitoring known persons in a small social network (Snell, 2023).

12. Virtual reality

Virtual reality environments are being used more frequently by neuroscientists to mimic social interactions and natural occurrences. VR generates interactive, multimodal sensory inputs that outperform other techniques for usage in applications and study in neuroscience. People's attention is diverted by petty stuff on social media and applications. As a result, the repercussions in the future will be even harsher. Online social interaction has surpassed offline social interaction in recent years (Devika & Oruganti, 2021).

13. Author's opinion and recommendations

Unnecessary busyness is also not good for the youngsters, so we should guide or mentor them for the right pathways so that they can become good social people and, when necessary, only then we have to attach with the social sites as compare to the society, so what we are going to give to the society at the end it will matter later on. It is our social responsibility that we have to come up with the solution to protect our societies from the social stigma and other things during odd situations so that there will be no issue in the future like problem or disorder due to this neuro-related problem (Garg & Gupta, 2022).

14. Outside of the lab

The field of social neuroscience is one that's both an expanding field of study with clear applications in areas other than academia, such as education, health, and public policy. As social neuroscience advances our understanding of human thought and our ability to predict, influence, and even control it, new ethical issues emerge. The investigation of the neurological correlates of dishonesty is one such example that has caught the attention of the criminal justice and intelligence agencies. The introduction of MRI scanners has renewed discussion regarding the sensitivity and reliability of such assessments, which has persisted for millennia in the hunt for an accurate and invariant brain signature for lying. Recent studies on the neurology of deception have brought up several significant issues, such as whether deception processes are particular or depend on a collection of general-purpose processes. It also brings up fresh questions regarding when, how, and if to protect one's own mental privacy. Our knowledge of agency—the consciousness of being in control of one's own acts as well as the outcomes of those actions on the outside world—is another instance of how neuroscience has affected moral, legal, and policy concerns) (Valliani et al., 2019). The concept of volition has long been used to judge a person's guilt in regard to criminal accusations since people are held accountable for their actions and the consequences of those actions in society. Recent findings in psychology, philosophy, neuroscience, and psychiatry have challenged the idea of free will as it pertains to human behavior. People may not be as accountable for their acts as the law presumes, according to certain theories, which have been sparked by the prediction of decisions, for instance, based on brain activities detected before the decision. If these considerations and conclusions are influenced by preceding brain activities, they are considered epiphenomenal (Orfanidis et al., 2022). Even if this were the case for just one individual, actions and interactions with other people are still the outcome of these considerations and judgments. Based on his or her antecedent brain functions, the individual's encounter with another person produces responses and a matching change in that person's brain function that is not totally foreseeable. The first person's statements and actions in the encounter have an impact on the second person's consciousness and underlying brain states, which in turn have an impact on the first person's consciousness and underlying brain states. The way something is communicated—verbally, nonverbally, or behaviorally—can have an impact on how other people's brains function. Thus, language and other deliberate behaviors can be crucial mediators between various brain states (Leo et al., 2022). This does not suggest that individuals have free will, but it does suggest that awareness (as a brain function) may act to influence one's (following) neural impulses through the effects on others. As decision-making, agency, and executive function neuroscience develop, it provides a more detailed and specific explanation of how the brain functions (Jude, 2021). We will have to rely on our hunches about individuals and what drives them. To emphasize the distinction, social neuroscience sees the brain as a mobile, broadband computer system that can link to and interact with other operating systems, in contrast to cognitive neuroscience, which sees the brain as an isolated computer. Both viewpoints are valid, of course, but they produce somewhat different issues, methods, conclusions, and viewpoints. In order to comprehend the brain from a social neuroscience perspective, we need to do more than just measure different levels of organization; we also need to try to define the mechanisms by which the processes we observe function, as well as the more fundamental principles that govern how the brain functions. The description and verification of these pathways and principles will advance with the use of both human and animal models, allowing for a deeper understanding of the complexity of social processes. This strategy, meanwhile, will also convey a faulty and unsophisticated knowledge of how social phenomena function in biological processes and how these processes can be scientifically explored. As a result, we think that accepting and enhancing the interdisciplinary structure of the field will increase the veracity of theories in social neuroscience as well as the area's transferability (Kidwai & Siddiqui, 2022).

15. Conclusions

Now a days, emotional things are moving away from societies, and they are focusing on individual growth rather than the growth of others, so these kinds of qualities can only be sustained with the help of social connections, so we must control the forceful implementation of automated technology because technology can make humans dependent, and if we become dependent on any technology, we will not be able to think properly. As a result, striking a balance between old and new technology is critical. People are now beginning to use automated diagnostic instruments; previously, biochemistry tests were performed manually.

We now have automated machines that give us results in fractions of seconds, diagnosis is quick, and we can save people's lives. Accepting new technology can be difficult, but if it makes people dependent, it is not a good thing. However, if it simplifies things and aids in problem solving, it is acceptable to accept the changes because they are necessary.

Acknowledgment

We would like to acknowledge the department of Medical Laboratory Technology, Chandigarh University for providing the required support.

References

Ansari, D., & Coch, D. (2006). Bridges over troubled waters: Education and cognitive neuroscience. *Trends in Cognitive Sciences, 10*(4), 146–151. https://doi.org/10.1016/j.tics.2006.02.007

Ansari, G., Garg, M., & Saxena, C. (2021). *Data augmentation for mental health classification on social media.* arXiv. https://arxiv.org.

Barrett, L. F., & Satpute, A. B. (2013). Large-scale brain networks in affective and social neuroscience: Towards an integrative functional architecture of the brain. *Current Opinion in Neurobiology, 23*(3), 361–372. https://doi.org/10.1016/j.conb.2012.12.012

Benton, S. L. (2010). Introduction to special issue: Brain research, learning, and motivation. *Contemporary Educational Psychology, 35*(2), 108–109. https://doi.org/10.1016/j.cedpsych.2010.04.007

Berntson, G. G., & Miller, G. A. (2018). John T. Cacioppo (1951–2018). *Psychophysiology, 55*(8). https://doi.org/10.1111/psyp.13200

Bruner, J. (1985). Models of the learner. *Educational Researcher, 14*(6), 5–8. https://doi.org/10.3102/0013189X014006005

Cacioppo, J. T., Ernst, J. M., Burleson, M. H., McClintock, M. K., Malarkey, W. B., Hawkley, L. C., Kowalewski, R. B., Paulsen, A., Hobson, J. A., Hugdahl, K., Spiegel, D., & Berntson, G. G. (2000). Lonely traits and concomitant physiological processes: The MacArthur social neuroscience studies. *International Journal of Psychophysiology, 35*(2–3), 143–154. https://doi.org/10.1016/S0167-8760(99)00049-5

Chiao, J. Y., Harada, T., Oby, E. R., Li, Z., Parrish, T., & Bridge, D. J. (2009). Neural representations of social status hierarchy in human inferior parietal cortex. *Neuropsychologia, 47*(2), 354–363. https://doi.org/10.1016/j.neuropsychologia.2008.09.023

Chiao, C., Lin, K. C., & Chyu, L. (2022). Perceived peer relationships in adolescence and loneliness in emerging adulthood and workplace contexts. *Frontiers in Psychology, 13.* https://doi.org/10.3389/fpsyg.2022.794826

Coch, D., & Ansari, D. (2009). Thinking about mechanisms is crucial to connecting neuroscience and education. *Cortex, 45*(4), 546–547. https://doi.org/10.1016/j.cortex.2008.06.001

Curley, J. P., & Ochsner, K. N. (2017). Neuroscience: Social networks in the brain. *Nature Human Behaviour, 1*(5). https://doi.org/10.1038/s41562-017-0104

Devika, K., & Oruganti, V. R. M. (2021). A machine learning approach for diagnosing neurological disorders using longitudinal resting-State fMRI. In *Proceedings of the confluence 2021: 11th international conference on cloud computing, data science and engineering* (pp. 494–499). Institute of Electrical and Electronics Engineers Inc. https://doi.org/10.1109/Confluence51648.2021.9377173

Eisenberger, N. I., & Cole, S. W. (2012). Social neuroscience and health: Neurophysiological mechanisms linking social ties with physical health. *Nature Neuroscience, 15*(5), 669–674. https://doi.org/10.1038/nn.3086

Ferrari, C., Lega, C., Vernice, M., Tamietto, M., Mende-Siedlecki, P., Vecchi, T., Todorov, A., & Cattaneo, Z. (2016). The dorsomedial prefrontal cortex plays a causal role in integrating social impressions from faces and verbal descriptions. *Cerebral Cortex, 26*(1), 156–165. https://doi.org/10.1093/cercor/bhu186

Garg, T., & Gupta, S. K. (2022). Efficient approaches to predict neurological disorder using social networking sites. In *Proceedings - 2022 IEEE 11th international conference on communication systems and network technologies* (pp. 294—298). Institute of Electrical and Electronics Engineers Inc. https://doi.org/10.1109/CSNT54456.2022.9787627

Glaser, R., & Kiecolt-Glaser, J. K. (1994). Stress-associated immune modulation and its implications for reactivation of latent herpesviruses. *Infectious Disease and Therapy Series, 13*, 245.

Goswami, U. (2009). Mind, brain, and literacy: Biomarkers as useable knowledge for education. *Mind, Brain, and Education, 3*(3), 176—184. https://doi.org/10.1111/j.1751-228X.2009.01068.x

Guo, Z., & Zhu, B. (2022). Does mobile internet use affect the loneliness of older Chinese adults? An instrumental variable quantile analysis. *International Journal of Environmental Research and Public Health, 19*(9). https://doi.org/10.3390/ijerph19095575

Hall, P. A., Rolls, E., & Berkman, E. (2023). The social neuroscience of eating: An introduction to the special issue. *Social Cognitive and Affective Neuroscience, 18*(1). https://doi.org/10.1093/scan/nsac060

Hammersley, C., Richardson, N., Meredith, D., Carroll, P., & McNamara, J. (2021). "That's me I am the farmer of the land": Exploring identities, masculinities, and health among male farmers' in Ireland. *American Journal of Men's Health, 15*(4). https://doi.org/10.1177/15579883211035241

Jude, H. D. (2021). Handbook of decision support systems for neurological disorders. In *Handbook of decision support systems for neurological disorders* (pp. 1—308). Elsevier. https://doi.org/10.1016/B978-0-12-822271-3.01001-X

Kedia, G., Harris, L., Lelieveld, G. J., & Van Dillen, L. (2017). From the brain to the field: The applications of social neuroscience to economics, health and law. *Brain Sciences, 7*(8). https://doi.org/10.3390/brainsci7080094

Kennedy, J., Frieden, L., & Dick-Mosher, J. (2021). Responding to the needs of people with disabilities in the Covid-19 pandemic: Community perspectives from centers for independent living. *Journal of Health Care for the Poor and Underserved, 32*(3), 1265—1275. https://doi.org/10.1353/HPU.2021.0130

Kidwai, M. S., & Siddiqui, M. M. (2022). Computer-based techniques for detecting the neurological disorders. In *EAI/Springer innovations in communication and computing* (pp. 185—205). Springer Science and Business Media Deutschland GmbH. https://doi.org/10.1007/978-3-030-77746-3_13

Larkin, K., Kawka, M., Noble, K., van Rensburg, H., Brodie, L., & Danaher, P. A. (2016). Empowering educators: Proven principles and successful strategies. In *Empowering educators: Proven principles and successful strategies* (pp. 1—167). Palgrave Macmillan. https://doi.org/10.1057/9781137515896

Leo, M., Bernava, G. M., Carcagnì, P., & Distante, C. (2022). Video-based automatic baby motion analysis for early neurological disorder diagnosis: State of the art and future directions. *Sensors, 22*(3). https://doi.org/10.3390/s22030866

Lewis, P. A., Rezaie, R., Brown, R., Roberts, N., & Dunbar, R. I. M. (2011). Ventromedial prefrontal volume predicts understanding of others and social network size. *NeuroImage, 57*(4), 1624—1629. https://doi.org/10.1016/j.neuroimage.2011.05.030

McIntyre, Jennifer (2015). *Neuroscientific possibilities for mainstream educators*. Springer Science and Business Media LLC. https://doi.org/10.1057/9781137515896_4

Meyer, M. L., & Lieberman, M. D. (2012). Social working memory: Neurocognitive networks and directions for future research. *Frontiers in Psychology, 3*. https://doi.org/10.3389/fpsyg.2012.00571

Meyer, M. L., Spunt, R. P., Berkman, E. T., Taylor, S. E., & Lieberman, M. D. (2012). Evidence for social working memory from a parametric functional MRI study. *Proceedings of the National Academy of Sciences of the United States of America, 109*(6), 1883—1888. https://doi.org/10.1073/pnas.1121077109

Michaels, J. L., Hao, F., Ritenour, N., & Aguilar, N. (2022). Belongingness is a mediating factor between religious Service attendance and reduced psychological distress during the COVID-19 pandemic. *Journal of Religion and Health, 61*(2), 1750–1764. https://doi.org/10.1007/s10943-021-01482-5

Mitchell, J. P., Macrae, C. N., & Banaji, M. R. (2005). Forming impressions of people versus inanimate objects: Social-cognitive processing in the medial prefrontal cortex. *NeuroImage, 26*(1), 251–257. https://doi.org/10.1016/j.neuroimage.2005.01.031

Murphy, P. K., & Benton Stephen, L.,S. L. (2010). The new frontier of educational neuropsychology: Unknown opportunities and unfulfilled hopes. *Contemporary Educational Psychology, 35*(2), 153–155. https://doi.org/10.1016/j.cedpsych.2010.04.006

Neuroscience: Social networks in the brain. (n.d.). Nature Human Behaviour.

Ochsner, K., Kruglanksi, A., & Higgins, E. (2007). *Social psychology: A handbook of basic principles*. Original work published 2007.

Orfanidis, C., Darwich, A. S., Cheong, R., & Fafoutis, X. (2022). Monitoring neurological disorders with AI-enabled wearable systems. In *DigiBiom 2022 - proceedings of the 2022 emerging devices for digital biomarkers* (pp. 24–28). Association for Computing Machinery, Inc. https://doi.org/10.1145/3539494.3542755

Parkinson, C., Kleinbaum, A. M., & Wheatley, T. (2017). Spontaneous neural encoding of social network position. *Nature Human Behaviour, 1*(5). https://doi.org/10.1038/s41562-017-0072

van Rensburg, H., & Adcock, B. (2016). Voices from Sudan: The use of electronic puzzles in an adult refugee community learning. In *Empowering educators: Proven principles and successful strategies* (pp. 155–166). Palgrave Macmillan. https://doi.org/10.1057/9781137515896

Sapolsky, R. M. (2005). The influence of social hierarchy on primate health. *Science, 308*(5722), 648–652. https://doi.org/10.1126/science.1106477

Shute, N. (2009). *Prescription: Don't be lonely* (Vol 146). US News World Rep. Original work published 2009.

Snell, J. (2023). Social neuroscience the rise of the behavorial sciences. *Education (Chula Vista), 143*, 44–47.

Spunt, R. P., & Lieberman, M. D. (2014). Automaticity, control, and the social brain. In J. W. Sherman, B. Gawronski, & Y. Trope (Eds.), *Dual-Process Theories of the Social Mind. 2* (pp. 279–299). New York: Guilford Press.

Steptoe, A., Owen, N., Kunz-Ebrecht, S. R., & Brydon, L. (2004). Loneliness and neuroendocrine, cardiovascular, and inflammatory stress responses in middle-aged men and women. *Psychoneuroendocrinology, 29*(5), 593–611. https://doi.org/10.1016/S0306-4530(03)00086-6

Tsakalidis, A., Chim, J., Bilal, I. M., Zirikly, A., Atzil-Slonim, D., Nanni, F., Resnik, P., Gaur, M., Roy, K., Inkster, B., Leintz, J., & Liakata, M. (2022). Overview of the CLPsych 2022 shared task: Capturing moments of change in longitudinal user posts. In *CLPsych 2022 - 8th workshop on computational linguistics and clinical psychology, proceedings* (pp. 184–198). Association for Computational Linguistics (ACL).

Valliani, A. A. A., Ranti, D., & Oermann, E. K. (2019). Deep learning and neurology: A systematic review. *Neurology and Therapy, 8*(2), 351–365. https://doi.org/10.1007/s40120-019-00153-8

Ward, J. (2012). *The student's guide to social neuroscience*. Original work published 2012.

Why loneliness is hazardous to your health - PubMed. (n.d.).

Zachariae, R. (2009). Psychoneuroimmunology: A bio-psycho-social approach to health and disease. *Scandinavian Journal of Psychology, 50*(6), 645–651. https://doi.org/10.1111/j.1467-9450.2009.00779.x

CHAPTER TWO

Social neuroscience: inferring mental states in social media

Umesh Gupta[1] and Vasu Jain[2]
[1]SR University, Warangal, Telangana, India
[2]SCSET, Bennett University, The Times of India Group, Greater Noida, Uttar Pradesh, India

1. Introduction

Social media has ingrained itself into everyday life and generated many data. As people increasingly use social media to express their emotions, thoughts, and opinions on several subjects, it has become a rich resource for researchers interested in psychology and human behavior. This has led to a rapidly growing field that applies machine learning (ML), artificial intelligence (AI), computer vision (CV), and natural language processing (NLP) techniques to predict mental states from social media posts, comments, and other data. This gives researchers a remarkable resource for studying human behavior and mental conditions. There has been a significant amount of research in this domain, and it has led to advances in the ability to detect sentiment, emotion, personality traits, and depression symptoms from data posted on social media. Research has shown that sentiment analysis algorithms can accurately detect the sentiment expressed in social media posts (Liu, 2012) and that emotion recognition algorithms can identify the emotions expressed in social media photos. Similarly, personality analysis algorithms can infer personality traits from profiles on social media (Choudhury, 2021), and depression detection algorithms can classify depression symptoms from social media data (Seabrook et al., 2018).

1.1 History

This interest in deducing mental states from social media platforms dates to the early 2010s. With the increase of online social networking platforms, such as MySpace in 2003, Facebook in 2004, and Twitter in 2006, researchers have begun to realize the promise of sourcing information from social media to understand human behavior and mental states. Some of the earliest research in this area was conducted by Liu (2012), who explored

Emotional AI and Human-AI Interactions in Social Networking
ISBN: 978-0-443-19096-4
https://doi.org/10.1016/B978-0-443-19096-4.00009-2

the uses of sentiment analysis algorithms to detect the sentiment expressed in social media posts, which laid the foundation for subsequent research in sentiment analysis, which has since become a widely studied area in the field of NLP. In the following years, researchers expanded the scope of their work to include other mental states and behaviors, such as emotions, personality traits, and depression. At the same time, (Choudhury, 2021) explored using social media data to predict depression. Deep recurrent neural networks are used to recognize emotions expressed in social media photos. The rise of deep learning and the availability of large-scale datasets derived from social media in recent years has led to significant advances in this field. For instance, (Coppersmith et al., 2014) used NLP techniques to identify depressive language on social media, while (Amanat et al., 2022) proposed a deep neural network model for detecting depression in online text and behavioral data. In conclusion, the history of predicting mental states from social media data is shaped by advances in NLP and AI technologies, as well as the growing recognition of the value of social media data for understanding human behavior and mental health.

1.2 Impact

The ability to infer mental states from social networks has profoundly impacted several sectors, including psychology, psychiatry, and healthcare. Using social media data to better understand human behavior and mental health can improve diagnosis and treatment outcomes while supporting population-level mental health and well-being research. The most significant impact areas include diagnosing and treating cognitive conditions such as depression and suicidal tendencies. For example, studies show that social networking data can be used to identify initial symptoms of depression, allowing for earlier intervention and potentially better outcomes.

Additionally, using social media data in mental health studies (Kaur et al., 2022) and research can help understand the causes and risk factors for mental health conditions while supporting the development of new treatments and interventions. Developing algorithms for sentiment analysis, emotion recognition, and personality prediction has pushed the boundaries of what is possible in these areas, leading to new and innovative approaches to understanding human behavior and mental states. Moreover, using social media data in AI research (Umesh Gupta & Gupta, 2022) has raised significant ethical and privacy concerns. Collecting and analyzing large-scale social media datasets raises questions about data privacy, informed consent, and the potential misuse of personal information. As such, it is vital for researchers in this field to consider these issues carefully and to develop responsible data practices that protect the privacy and well-being of individuals. To summarize, the impact of inferring mental states in social media has been significant, with potential benefits for mental health research and treatment and the advancement of AI and NLP technologies. However, it is crucial to carefully consider this work's ethical and privacy implications and develop responsible data practices to protect individuals and communities.

2. Related work

The academic chapter (Choudhury, 2021) comprehensively outlines the techniques to predict mental health status from desired social media data. The authors discuss the strengths and limitations of different approaches, including NLP and sentiment analysis, CV, multimedia fusion, and ML algorithms. They also highlight the need for further research to improve these techniques' accuracy and robustness and address ethical and privacy concerns associated with inferring mental states using social networking data. To uncover depression signals that foretold the development of depressive behaviors, authors (Choudhury, 2021) crowdsourced depression screening questionnaires and users' Twitter posts and feeds and analyzed their language and stated emotions for network activity. This research (Moreno et al., 2011) found that Facebook status posts by college students can disclose signs of significant depression episodes. To find indicators of shifts to suicidal thoughts in subreddits discussing mental health and suicide support, authors (De Choudhury et al., 2016) examined the linguistic structure, its matching styles, and interaction patterns of users on Reddit. In Instagram user activity, lexical, behavioral, and thematic changes were examined in postings on eating disorders (Reece & Danforth, 2017). Another study examined Internet usage patterns, including average packets-per-flow, text octets, and Internet packets, and discovered a correlation between these patterns and depressive symptoms (Katikalapudi et al., 2012).

Studies have more recently added visual content and other modalities that can be used to evaluate mental health states. To uncover visual cues that may be combined with linguistic and user activity features to predict signs of depression in Instagram postings (Reece & Danforth, 2017), performed face detection and color analysis on user-generated photographs. Research analyzing user dimensions from language and visual data, such as sentiment, emotion, personality, and demographics, is also relevant to this field of study. The identification of sentiment in social media photographs has been addressed by authors (Yilin & Baoxin, 2015) using visual and contextual network data, such as comments by friends and users' descriptions. The research published by Wendlandt et al. (2017) investigated how user-generated photos and their descriptions may be used to predict gender and personality. This study provided correlational methods to investigate the predictive value of various visual and textual variables in predicting user attributes. It was shown that these methods perform better than models based on only one modality at a time. An extensive collection of visual, linguistic, and postmeta characteristics generated from Flickr postings are analyzed in work using a similar method to determine the health state of individuals.

2.1 Ethical dilemma

The use of social media data in inferring mental states has produced several severe ethical concerns and quandaries. The authors (Chancellor et al., 2019) discuss many significant ethical conflicts when using social media data. One of the main ethical dilemmas is related to privacy and data security. Collecting and analyzing large-scale social media datasets raise questions about individuals' privacy and the potential misuse of personal information. There is also a risk that sensitive information about individuals' mental health could be disclosed to third parties or used for purposes other than those intended by the individuals who shared the news. Another ethical tension described in the chapter is related to informed consent. Social media users may be unaware of their data or the repercussions of such use, leading to questions about whether their support is fully informed.

Additionally, there is a risk that social media data could reinforce existing power imbalances and lead to discrimination against certain groups. Using social media data to determine mental states also raises questions about the accuracy and validity of the results. There is a risk that algorithms and models used for this purpose could produce biased or inaccurate results, particularly in cases where data from certain groups of people are underrepresented or poorly understood. Finally, there is also a risk that using social media information to infer mental states could lead to unintended consequences, such as stigma and discrimination. For example, individuals identified as having a mental health condition based on their social media data may face discrimination in workplaces or in their personal lives. The ethical dilemma of deducing mental states from social media is complex and multifaceted. It raises questions about privacy, informed consent, accuracy, and unintended consequences. It is crucial for researchers and practitioners in this field to consider these ethical concerns carefully and to develop responsible data practices that prioritize the privacy and well-being of individuals.

2.2 Effect of social media on mental health

Social media's effect on mental health is a growing concern among researchers and the public. Although social media can provide support, increase social connectedness, and offer access to information and resources about mental health. It also has been linked to several adverse effects, including cyberbullying, exposure to unpleasant or triggering content, and reduced face-to-face social contact. Numerous studies have investigated the correlation between social media use and mental health effects, including depression, anxiety, and body dissatisfaction. For instance, a study by Rosen et al. (2013) discovered that increased interaction with social media was associated with expanded indications of depression in young adults. Another (Kross et al., 2013) survey showed that Facebook usage was positively related to signs of depression, especially among individuals who compared their lives to others on the platform.

It should be noted, however, that the relationship between social media and mental health is complex and bidirectional. For example, individuals with pre-existing mental health illnesses may be more inclined to use social media, and increased social media usage may intensify pre-existing mental health troubles. Furthermore, the type and frequency of social media use can significantly impact mental health. For example, passive scrolling through social media feeds has been associated with adverse well-being effects. In contrast, active engagement in social media activities, such as communication with friends, has been associated with positive developments.

To summarize, the impact of social media on mental health is a complicated and intricate issue that requires further investigation. Even though social media can offer support and increase social connectedness, it can also adversely affect mental health, especially among vulnerable individuals. As such, it is essential to consider the context and nature of social media use when examining its impact on one's mental health.

2.3 Technologies and methodologies developed

The technology for inferring mental states in social media has rapidly advanced in recent years, developing new methodologies and approaches. In the academic chapter by Xu et al. (2020), the authors describe the use of multimodal information to tackle this problem. One approach used is sentiment analysis and NLP. This entails evaluating the text of social media posts to identify language patterns indicative of mental health problems such as sadness or anxiety. For example, individuals experiencing depression may use more negative or self-referential language in their posts, which can be detected using NLP and sentiment analysis algorithms. Another approach in this field is a CV, which involves analyzing images and videos posted on social media to infer mental health status. For example, researchers have used CV algorithms to interpret body language and facial expressions to deduce the emotional state of individuals in photos or videos. The authors also describe multimedia fusion techniques to combine information from multiple modalities, such as text, images, and videos, to infer mental health status. This approach takes advantage of the complementary strengths of different modalities and can lead to more accurate and robust results. Finally, the authors describe the use of ML algorithms to implement predictive models, which can be utilized for inference from social media data (Xu et al., 2020). These models can be trained on large datasets of posts associated with mental health labels and then be used to predict the status of new users' mental health.

2.4 Datasets

We have explored the datasets related to inferring mental states in social media in Table 2.1.

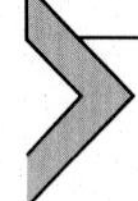

3. New techniques for understanding emotions in social media

It is challenging to build a corpus of literary reviews that span disciplinary boundaries due to publishing practices. In contrast to other disciplines that rely on journals, conference proceedings are the most popular venues for publication in computer science. Journal entries were thoroughly indexed when evaluating original search techniques using conventional indexing services. However, conferences widely acknowledged as significant in these subdomains by professional associations had substantial gaps.

Table 2.1 Different dataset repositories related to inferring mental states in social media.

Dataset	Content	Source
Social media data for sentiment analysis (Choudhary, 2017)	Facebook comments, Tweets, and Amazon reviews	http://cucis.ece.northwestern.edu/projects/Social/sentiment_data.html
Dreaddit (Turcan, 2019)	190K posts from Reddit communities	https://paperswithcode.com/dataset/dreaddit
Social media and mental health (Braghieri, 2022)	Data and code from several sources	www.openicpsr.org/openicpsr/project/175582/version/V1/view/
Mental health datasets (Harrigian et al., 2021)	A collection of research articles	https://paperswithcode.com/paper/on-the-state-of-social-media-data-for-mental

Initial keyword search efforts using tools like Google Scholar produced more than 200,000 candidate publications that are unsearchable. This search was "seeded" by 41 hand-selected venues to address these difficulties. Then, candidate articles are filtered for in these venues using search phrases. Finally, some candidates' referrals are randomly selected to find any missing research.

3.1 Search strategy

Two sets of keywords were created for pair-wise searches: one for social media and the other for mental health. There were 16 synonyms for mental health, including but not limited to symptomatology, most prevalent mood and psychosocial illnesses, and generic terminology for mental health and disorders (e.g., stress, psychosis) (Chancellor & De Choudhury, 2020). The DSM-V101 and earlier research 20 and 21 served as sources for this. Eight terms related to social media were searched, including the general terms for the three most well-known social networks: Facebook, Twitter, and Instagram. Table 2.2 (Chancellor & De Choudhury, 2020) contains a list of keywords. To overcome the abovementioned indexing issues, 41 English sites that might publish studies on MHS prediction using social media data were found.

Regarding the concerns about indexing, three different search engines were used to confirm comprehensive analysis across various venues. The ACM Digital Library (ACM Digital Library. *Retrieved from ACM Digital*

Table 2.2 Literature search: Keywords.

Group	**Keywords**
Mental health	Mental health, mental disorders, mental wellness, posttraumatic stress, schizophrenia, borderline personality disorder, bulimia, stress, depression, obsessive-compulsive disorder, psychosis, anxiety, suicide, eating disorder, disorder, anorexia, bipolar disorder
Social media	Social media, social networking sites, social networks, SNS, Facebook, Twitter, Instagram, forum
Search term	Keywords in Mental Health and Social Media Group

From Chancellor, S. & De Choudhury, M. (2020). Methods in predictive techniques for mental health status on social media: a critical review. *Npj Digital Medicine, 3*(1). https://doi.org/10.1038/s41746-020-0233-7

Library, 2023) was used for journals from ACM conferences. Google Scholar with the Publish or Perish service was utilized for additional conference publications, and Web of Science was used for journals. This includes a collection of proceedings from data science and health informatics conferences and general interest journals. The conference proceedings were manually scanned for titles and abstracts containing relevant keywords, as search engines indexed one venue (Tsakalidis et al., 2022) inappropriately. By using these techniques, 4420 texts were found that corresponded to keyword pairs.

3.2 Filtering strategy

Only original, peer-reviewed, and archived works released from 2008 to 2017 were allowed to remain in the manuscripts, coinciding with the beginning of the academic study on social media. Since they did not meet the criteria for originality, certain publications were disqualified, including meta and literature reviews, reflections and thoughts, collective tasks, case studies, and unarchived submissions to computer science conferences. This produced 2344 texts after deduplication and filtering. Next, results blatantly unrelated to mental health or social media were removed by filtering by title and abstract. Other medical problems, including cancer, and data sources like electronic health recordings are a few examples of mismatches. Eighty-seven articles were chosen from the titles and abstracts that were screened.

3.3 Analysis technique

By following the work (Chancellor & De Choudhury, 2020) and understanding of the study field, an analytical rubric comprising other articles' descriptive, quantitative, and qualitative criteria. This rubric has over 100 criteria addressed, including data collecting and data preprocessing methodologies, accuracy and benchmarks, results presentation systems, and the availability of comments on specific study design decisions and the research's implications. In addition, qualitative notes were recorded for desirable thematic observations and analytical insights. Four random texts from the corpus were chosen to annotate and verify the robustness of the rubric. Based on the outcomes of trial annotations, the title for additional reporting areas was modified. The Supplementary Information, Table 2.2, contains the pertinent elements of the rubric design. The 75 articles in the corpus were then all closely examined, with the rubric annotated to reveal patterns across the corpus. To standardize the coding procedure, the first author reviewed and coded the complete dataset twice, in random order. Then, the analysis is concluded by reviewing the emerging themes and data.

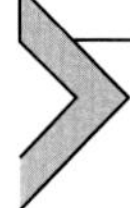

4. The power of multimodal data analysis in mental state inference

Multimodal data analysis takes advantage of the complementary strengths of various data sources, leading to more accurate and robust results. For instance, text data can provide valuable insights into an individual's thoughts and feelings. In contrast, image and video data can provide information about facial expressions, body language, and other nonverbal cues. Combining information from these various data sources can lead to a more comprehensive understanding of an individual's mental state. The authors (Xu et al., 2020) propose a multimodal fusion approach to infer mental health status from individuals' social media data. This approach involves utilizing various data analysis techniques to extract information from each data source and then fusing the information to make a final prediction. The authors found that this approach outperformed traditional single-modality methods like NLP and sentiment analysis. Another benefit of multimodal data analysis is that it can supply a more nuanced understanding of mental health status. For example, text data may indicate an individual is experiencing depression, but image and video data may show that they can still smile and engage with others. This information can paint a complete picture of a person's mental state, informing care and supporting decisions. The power of multimodal data analysis in mental state inference has been demonstrated in numerous studies (Xu et al., 2020). This approach takes advantage of the complementary strengths of various data sources, leading to more accurate and robust results and a more nuanced understanding of mental health status.

4.1 Preprocessing and experimental setup

During the experiments, 10-fold cross-validation is used to conduct assessments at the user level. To perform the tests, a balanced number of posts (11,828) is obtained from each set of users randomly dropping the extra seats. All features are adjusted to zero mean and unit variance. As a result, identical user posts appear in the training or the test sets. The classifiers: adaptive boost (AB), decision tree (DT), and support vector machine with a linear kernel (SVMLinear) are employed. A two-layer neural network classifier built with the TensorFlow toolkit (Abadi, 2015) is evaluated. The activation function for the hidden layers is a rectified linear unit (ReLU), and the output probability is restricted between zero and one by a sigmoid function. The binary classification is then performed using a numerical threshold, usually 0.5. A channel valve is added to experiment with different feature combinations. The default parameters for the SVM and adaptive boost classifiers are used in sci-kit-learn. From the first concealed layer, each feature set's use is regulated. As a result, when the valve is turned on, its output equals the input and is zero otherwise.

4.2 Predicting mental health status using multimodal observations

Many tests utilized the feature sets to independently and together assess the predictive potential of visual, language, and metadata information. The features are pulled from either image (optical), captions (linguistic), or postmeta elements (meta), as well as mixtures of them by simply concatenating the various feature vectors. Fig. 2.1 displays classification results for healthy users and posts generated by people with a mental disease. Fig. 2.2 shows classification results for healthy users and those at risk of mental illness. Overall, we see that integrating all modalities simultaneously outperformed using single modalities for most classifiers. This implies that combining visual, verbal, and post information can enhance the accuracy of a user's mental health condition prediction. The multimodal approach was more successful when

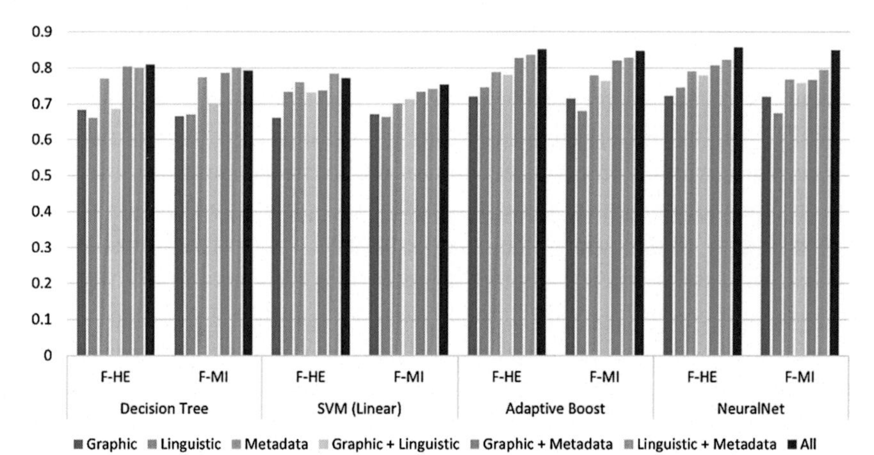

Figure 2.1 ***F1-scores for classification of healthy (HE) and mentally ill (MI) users utilizing four algorithms with visual, verbal, and postmeta variables (meta)*** Bar chart of classification of healthy and mentally ill users over F1-score. *Data from Xu, Z., Pérez-Rosas, V., & Mihalcea, R. (2020). Inferring social media users' mental health status from multimodal information. In LREC 2020—12th International Conference on Language Resources and Evaluation, Conference Proceedings (pp. 6292–6299). European Language Resources Association (ELRA).*

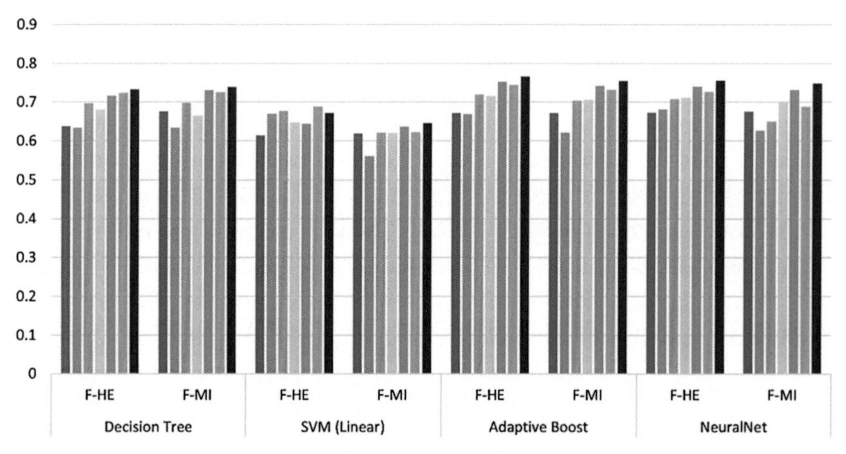

Figure 2.2 ***F1-scores for classification of posts from healthy users (HE) and posts from users at risk of mental illness (PREMI) utilizing four algorithms and visual, language, and postmeta features (meta).*** Bar chart for classification of healthy (HE) and risk of mental illness (PREMI) over F1-scores. *From Xu, Z., Pérez-Rosas, V., & Mihalcea, R. (2020). Inferring social media users' mental health status from multimodal information. In LREC 2020—12th International Conference on Language Resources and Evaluation, Conference Proceedings (pp. 6292–6299). European Language Resources Association (ELRA).*

comparing posts from the mental illness and illness-free control groups regarding task performance. The fact that users' behaviors in the present illness sets are more comparable to those in the control group may make it challenging to distinguish between the two groups clearly. But improved performances can be observed when combining the various feature sets between posts from before a mental illness and posts from healthy posts. Regarding the classification approach, the neural classifiers that utilize all features archive the best F-HE and FMI scores. It is curious to note that using language characteristics results in the lowest F-HE and FMI scores for neural classifiers, as shown in Figs. 2.1 and 2.2. The lack of linguistic feature vectors may have led to this. Future efforts will attempt to replace hand-engineered features with vector representations to overcome this issue.

5. The role of natural language processing

Natural language processing is a specialization of AI from its humble roots as a field dedicated to translating enemy messages during World War II to now managing several much more complex tasks such as machine translation, summarization, entity recognition, question answering, semantic search and analysis, etc., the field of NLP has seen some of the quickest growth among the various disciplines under AI. Since several assessments regarding emotion, sentiment, and mental states from social media use text in forms such as text posts, image captions, and comments, advancements in NLP play a vital role in the study in this field. Current model methods utilize numerous deep learning systems such as convolutional neural networks (CNNs), long short-term memory (LSTMs), and newer transformer models (such as Google's BERT). Data are cleansed and preprocessed before being used to train and predict models. In Chen et al. (2023), the data were cleaned by removing random characters/text artifacts (also known as noise) and removing emojis, URLs, and non-ASCII characters for easier processing. There is an understanding that even though some information may be lost by doing this, model performance is generally better. Other preprocessing techniques include converting all text letters to lowercase to increase uniformity and replacing all contractions with their unabbreviated forms to remove their ambiguity. The final preprocessing step involves tokenization, which breaks down large bodies of text into smaller chunks for individual processing. These could be words, phrases, punctuations, or even entire sentences. More techniques include scaling and normalizing numerical data, as shown in Eq. (2.1). Numerical data can have varying ranges; normalizing them in a fixed range can help models learn much faster.

$$Open\ \underline{value} = \frac{value - min(value)}{max(value - min(value)}\tag{2.1}$$

Eq. (2.1): *Normalization formula.*

In recent years, breakthroughs in NLP have involved turning words into numerical vectors to make mathematics based around words easier. Several approaches like autoencoders, word2vec, and doc2vec have been used for quite some time. However, since the introduction of Transformers in 2017, these approaches have been directly using or tuning pretrained models trained on large amounts of data. These numerical vectors taken from words are known as "embeddings," and (Chen et al., 2023) used Sentence BERT to obtain the embeddings for its classification, using these vectors to create a matrix for embeddings for all users under study. CNNs are artificial neural networks inspired by the structure and process of biological visual cortexes. CNN is a multilayered architecture, with the initial layers focusing on low-level feature extraction from raw data and the last layers focusing on high-level feature extraction based on all layers before them. Fig. 2.3 below shows the general structure of the CNN used.

Since inference of mental states, emotions, mental illness symptoms, etc., are classification-based tasks, the metrics used to evaluate and evaluate the model's performance are primarily based on metrics such as accuracy, precision, recall, and F1-score. While accuracy is regarded as the best "at a glance" comparison, all the other metrics play key roles in identifying the weak spots of any classifier. The following equations are used to generate the metrics: TP is the true-positive value, TN is true-negative value, FP is false-positive value, FN is the false-negative value, and TPR is the true-positive rate, as shown in Eqs. 2.2–2.5.

$$Accuracy = \frac{TP + TN}{TP + TN + FP + FN}\tag{2.2}$$

Eq. (2.2): *Accuracy formula*

$$Precision = \frac{TP}{TP + FP}\tag{2.3}$$

Eq. (2.3): *Precision formula*

$$Recall = TPR = \frac{TP}{TP + FN}\tag{2.4}$$

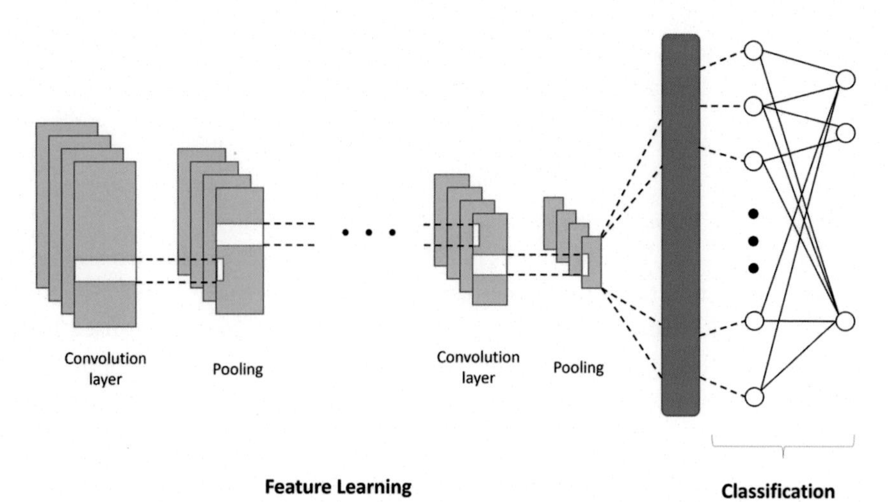

Figure 2.3 ***Convolutional neural network (CNN).*** *From Chen, Z., Yang, R., Fu, S., Zong, N., Liu, H., & Huang, M. (2023). Detecting Reddit Users with Depression Using a Hybrid Neural Network. arXiv. https://doi.org/10.48550/arXiv.2302.02759 Convolutional Neural Network (CNN).*

Eq. (2.4): *Recall formula*

$$F1 = 2 \times \frac{Precision \times Recall}{Precision + Recall} \tag{2.5}$$

Eq. (2.5): *F1-score formula.*

Another technique primarily regarded as outdated but noteworthy nonetheless is a component of speech tagging (POS tagging). As highlighted in Burnap et al. (2015), part-of-speech tagging as feature extraction led to not insignificant results for the analysis of suicide-related communication over Twitter. Methodologies based on POS Tagging vary greatly, but the general approach involves making a feature set based on segregating words into their corresponding grammatical parts of speech. This could include nouns (possibly further classified into proper, common, plural, etc.), verbs (further classified based on tense), and others such as pronouns, conjunctions, determiners, symbols, numbers, etc. Processes involving POS Tagging are not statistically sound compared to their deep learning counterparts performing similar actions, but (Burnap et al., 2015) managed to obtain an F1-Score of 0.69, which, while not ideal, displays promise.

5.1 The use of computer vision for mental state analysis

CV is vital in multimodal applications based on social media. Xu et al. (2020) uses Flickr posts to gauge its users' mental health, a website primarily focused on video and image hosting. CV is vital in multimodal applications based on social media. Xu et al. (2020) uses Flickr posts to gauge its users' mental health, a website primarily focused on video and image hosting. The leading deep learning technology behind almost all CV and analysis remains CNN (Fig. 2.3) since all images are essentially a matrix of several values that indicate color, brightness, transparency, etc.

Xu et al. (2020) uses low-level feature extraction from posts such as color, texture, brightness, faces, objects, and scenes. All these factors, they claim, can be used to figure out the poster's mental state. Their study concluded that persons experiencing mental illnesses and those susceptible to mental illnesses tend to post darker or gloomier pictures with high contrasts showing indoor scenes and fewer facial images than healthy users. However, their F1 scores range from 0.75 to 0.85, and the claim and the statistics behind them could be more specific.

6. Challenges and limitations

Due to various limits and problems, detecting mental states via social media is difficult due to different limits and issues. Multiple academic articles have discussed these limitations and challenges, including ethical tensions research by Chancellor et al. (2019) and predictive techniques for mental health status by Chancellor and De Choudhury (2020). One major challenge in detecting mental states from social media data is inaccuracy. Social media data are often noisy, inconsistent, and unstructured, making it challenging to detect mental states accurately (Courville, 2016). Preprocessing this data can also lead to the loss of valuable information, making it difficult to achieve high evaluation results. Another challenge is the issue of privacy and ethics. Using social media data to infer mental states raises essential privacy and ethical concerns, such as the potential misuse of personal information and the infringement of an individual's right to privacy. This needs to be addressed more. Still, recent efforts by the European Union to further privacy indicate that it needs to be more on the radar of current government policymakers and officials. Many of these technologies and methodologies are currently in their initial stages of development and have yet to be thoroughly tested or validated. There is also a need for more standardization in the methods applied to detect mental states, leading to inconsistencies at worst and a lack of comparability between studies at best. In addition, the interpretation of mental states from social media data is affected by cultural and individual differences, such as the use of different languages and cultural norms. This can result in a lack of generalizability and a need for culturally sensitive approaches to detecting mental states in social media. However, it is nearly impossible to eliminate this issue in the current world. Finally, social media users may not be conscious of their data being used to infer mental states and may not have consented. It is vital to ensure that social media users know how their data are being used and to obtain support for any use of their data for mental state inference. These challenges need addressing to develop ideal approaches to detecting mental states from social media data. However, sufficient workarounds have been implemented to solve the graver issues, such as suicide prevention, already implemented on Meta websites and Google.

6.1 Overcoming the limitations of annotated data

Annotated data is a crucial component of ML algorithms, as it provides the necessary information for the algorithm to learn and make accurate predictions. However, annotated data also have several limitations that can impact the performance of the algorithms. Some of the most significant limitations include (Courville, 2016).

- Data availability: Annotated data are often limited, especially in specialized domains, leading to data scarcity, which can result in biased or inaccurate models.
- Data quality: The quality of annotated data is another major limitation, as errors or inconsistencies in the annotations can impact the performance of the algorithms.
- Data imbalance refers to the unequal proportion of samples in different categories, leading to biased models favoring the majority class.
- Annotation cost: The cost of annotating data is often high in terms of time and funds, limiting the extent of annotated data available.

Several strategies have been suggested to overcome these restrictions, including data augmentation (Ansari et al., 2021), transfer, active, and unsupervised learning (Settles, 2009). Data augmentation involves creating synthetic samples from existing data to increase available annotated data. This can be accomplished by adjusting existing images, such as rotation, scaling, or flipping, or by producing new data based on current data. Transfer learning involves shifting knowledge from one task or domain to another. This can be done by fine-tuning a pretrained model or using transfer learning to leverage understanding from related jobs or environments. Active learning involves humans in selecting which examples should be annotated based on the algorithm's performance. This can substantially reduce the volume of annotated data that are needed, as well as improve the quality of the annotations. Unsupervised learning involves training algorithms without annotated data using clustering for dimensionality reduction techniques. This can be useful when annotated data are unavailable or are too costly to obtain. The limitations of annotated data can significantly impact the performance of ML algorithms. However, these limitations can be overcome by applying data augmentation, transfer learning, active learning, and unsupervised learning, leading to more accurate and reliable algorithms.

6.2 Importance of interdisciplinary collaborations

Several previous studies have recognized the need for interdisciplinary collaboration to achieve an effective and accurate result (Xu et al., 2020; Chancellor & De Choudhury, 2020) provide a scathing criticism of some blatant issues that implementing good collaboration can solve. One of the significant advantages of interdisciplinary collaboration is the capability to bring together different areas of expertise to address the complex and multifaceted problem of mental state inference from social media. Computer scientists, psychologists, and sociologists can work together to develop algorithms that accurately detect mental states while considering the social and cultural context of social media use. Interdisciplinary collaboration also addresses ethical and privacy concerns comprehensively. Collaborating with experts in ethics and confidentiality can help ensure that the development of mental state inference algorithms is guided by moral and confidentiality considerations, thereby reducing the risk of privacy violations or unethical practices.

Additionally, interdisciplinary collaboration can provide a more thorough understanding of the complex interplay between mental states, social media use, and other influences, such as demographic characteristics and cultural norms. This can help to develop more accurate and practical algorithms for detecting mental states from social media data. Finally, interdisciplinary collaboration can provide opportunities for cross-disciplinary knowledge exchange and cooperation, leading to new theories, methods, and technologies for detecting mental states from social media. This can have far-reaching impacts, not only in the mental health field but also in related fields, for instance, psychology, sociology, and computer science. For example, in ASD disorder, also known as a behavioral disorder, the issue begins in infancy. It affects more habits as the child ages, moving to adulthood and beyond. Based on freely released UCI datasets, the assessment potential of the ELM and RVFL models with various optimization functional units is explored to measure the autism spectrum disorder in a human group that includes kids, young teenagers, and grown-ups (Gupta et al., 2022). This will be further explored in inferencing by analyzing mood switches. By bringing together experts from different fields and disciplines, interdisciplinary collaboration can address the complex and multifaceted problem of mental state inference from social media while considering ethical and privacy considerations, providing a comprehensive understanding of the problem, and promoting cross-disciplinary knowledge exchange and collaboration.

6.3 Integrating mental state inference with other social media analysis techniques

Integrating mental state inference with other social media analysis techniques has become an important study area in recent years. Inferring mental states using social media can provide a valuable understanding of individuals' thoughts, feelings, and emotions. By combining these insights with other forms of social media analysis, researchers can gain a more comprehensive understanding of individual behavior and mental well-being. Integrating mental state inference with network analysis techniques allows for examining connections between individuals on social media. Researchers can identify patterns of behavior that may indicate mental health. For instance, people with strong social relationships may be less prone to experience mental health issues, while those with weaker connections may be more susceptible (Yazdavar et al., 2020). In addition, researchers are exploring the integration of mental state inference with ML techniques. ML algorithms can be trained on vast amounts of data to recognize patterns that indicate mental states. Researchers can achieve even more accurate results by integrating this technology with mental state inference. Combining mental state inference with other social media analysis techniques can provide a more thorough understanding of individuals' mental well-being and behavior on social media. Further research is needed to realize this approach's potential fully.

7. Applications of inferencing

Detecting mental states from social media has many potential applications, including interventions, monitoring social media, and marketing research. This has been explored in several academic articles (Xu et al., 2020; Chancellor et al., 2019). Early detection of mental health disorders is one of the social media's most common applications of mental state inference. Algorithms can detect indicators of mental health conditions such as depression, anxiety, or stress by evaluating the content and actions of social media users. This can lead to earlier and more effective counseling and increase the effectiveness of these services. Another application of mental state inference in social media is social media monitoring. By analyzing the mental states of social media users, organizations can monitor and understand the mood, opinions, and behavior of individuals, groups, and communities on social media. This can provide a valuable understanding of individuals' opinions and emotions, helping organizations make informed decisions, such as product development, marketing strategies, and public relations. Mental state inference in social media can also have applications in marketing research. By analyzing the mental states of social media users, researchers can gain insights into the opinions and actions of individuals toward certain products, services, or brands. This can provide valuable information for product development, marketing strategies, and customer engagement. In conclusion, inference of mental states has a broad range of potential applications, including early detection of social media monitoring, mental health problems, marketing research, and the development of personalized mental health interventions. The result of accurate and practical algorithms for detecting mental states from social media data has the potential to have far-reaching impacts, not only in mental health but also in related fields, such as marketing, sociology, and computer science.

7.1 Social media and its role in mental health diagnosis and treatment

The part that social networks can play in mental health diagnosis and therapy is gaining traction. Several academic studies, such as Garg (2023) and Tsakalidis et al. (2022), have investigated the feasibility of using social media data to infer mental health status and moments of change and advise healthcare interventions. However, the topic of tightly integrating social media and its users' mental health has remained complex and contentious. Several academic articles have explored using social media data to diagnose mental health disorders and provide appropriate treatment. For example, a study by Safa et al. (2022) investigated using Twitter data to detect depression in adolescents. The authors concluded that social media data might be utilized to accurately predict depression in adolescents, demonstrating the utility of social media data in mental health diagnosis. Another study (Choudhury, 2021) investigated using social media data to assess depression symptoms in real-time. The authors discovered that social media data might be used to follow changes in depression symptoms over time, lending credence to using social media data in mental health treatment. Despite these promising findings, integrating social media data into mental health diagnosis and treatment raises significant ethical and practical concerns. For example, issues around data privacy and the accuracy of mental health diagnoses based on social media data are still significant challenges. The role of social media in mental health diagnosis and therapy is a complex and quickly growing topic. More research is needed to fully understand the potential and limitations of social media data in this context and develop effective and ethical methods for combining social media data into mental health diagnosis and treatment.

7.2 Personality profiling

The study of inference from social media has gained significant attention and popularity due to the abundance of personal information and data that can be obtained from social media platforms. Personality profiling, which involves the extraction of traits related to an individual's psychological characteristics, has been identified as a potential application of mental state inference. In recent years, researchers have explored the feasibility of inferring personality traits from social media data, focusing on NLP techniques. Past research has shown that mental state inference from social media can supply valuable insights into an individual's personality traits. A study (Choudhury, 2021) demonstrated the ability of ML algorithms to predict the Big-Five personality types of individuals centered on their social media text posts. Another study by Ferwerda et al. (2016) delves into using social media data for a given user, such as the number of friends, likes, and posts, to predict personality traits. These findings show that inferring a user's mental health condition can be helpful in personality profiling. However, the field is still relatively young, and several challenges must be explored and addressed before it can be widely used for personality profiling. The reliability of the inferences made from social media data is still a concern. In addition, the annotation of the data used for training the algorithms is often a challenge due to the subjectivity and variability of the data. Despite these challenges, the potential of mental state inference from social media for personality profiling is undeniable. Further exploration and research are needed to address these challenges and explore this technology's potential for various applications, including but not limited to mental health disorder diagnosis and treatment.

8. The need for cultural and contextual awareness

The process of inferring mental states from social media involves analyzing text, images, and other data sources to determine a person's emotional, psychological, and behavioral patterns. However, this process must be approached with cultural and contextual awareness, as these factors can significantly impact the accuracy of mental state inference. For example, a study (Imran et al., 2020) found that cultural differences can affect the portrayal of language and the degree or nature of emotion expressed in social media posts, which can hinder the accuracy of mental health status inference algorithms. Another study (Deng & Yang, 2021) discussed how cultural context could shape the expressions of emotion online. This could also confuse algorithms and cause their reliability to decline. Additionally, socio-political factors can impact the interpretation of mental states in social media. For example, a study (Sandoval-Almazan & Valle-Cruz, 2020) found that the language used in social media posts can be influenced by the political climate, which can impact the accuracy of mental state inference algorithms. Given these considerations, mental state inference algorithms need to be developed with an awareness of cultural and contextual factors to ensure their accuracy and reliability. This requires interdisciplinary collaboration between experts in computer science, psychology, sociology, and other relevant fields, to develop a comprehensive understanding of the impact of these factors on mental state inference from social media.

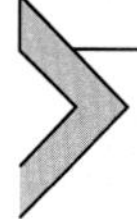

9. A comprehensive review of mental state inference in social media: Current state and future directions

Due to the increasing incidence of mental health concerns and the expanding influence of social networking sites, mental state inference from social media has received substantial interest from researchers and practitioners in recent years. Inferring mental health statuses from social media is a complex and multidisciplinary effort that requires expertise from several domains, including computer technology, psychology, sociology, and neuroscience. The development of mental state inference techniques has been motivated by the need to understand social media users' mental well-being better and help prevent and treat mental health issues. To this end, several studies have been conducted to explore using various data sources, including text, images, video, and audio. One of the most pressing challenges in using ML for mental state inference is the need for a clear definition of mental states and the absence of ground truth data. Additionally, the complex nature of mental health and well-being and the limitations of existing data sources and analysis techniques pose significant challenges for mental state inference. Despite these challenges, recent research has shown promising results in mental state inference from social media, using modern ML techniques, such as CNN, transformers, and multimodal data analysis. Multimodal data effectively improve mental state inference quality, considering multiple aspects of an individual's behavior and expression. Moreover, interdisciplinary collaboration between researchers and practitioners from different fields is essential for advancing mental state inference from social media. Such collaboration can bring together diverse expertise and perspectives, leading to more comprehensive and practical solutions. In conclusion, mental state inference from social media is a rapidly evolving field with numerous applications and tremendous potential to improve the understanding of well-being and how ML can help. However, this task's challenges and limitations pose a challenge too great to be overcome by current means. Research is needed to develop more accurate and reliable mental state inference solutions.

10. Conclusion and future Scope

The future of mental state inference from social media data is rapidly growing in importance with the potential to revolutionize mental health diagnosis and treatment. As the popularity of social media among the general public rises, the data available for analysis are growing exponentially. This presents opportunities and challenges for researchers and practitioners who aim to infer mental states from social media. Academic studies have shown the potential of mental state inference to accurately predict various mental illnesses, such as depression, using multiple methods. However, the current limitations of mental state inference include the need for annotated data, cultural and contextual awareness, and the limits of existing methods.

Researchers have explored interdisciplinary collaboration to overcome these limitations, integrating mental state inference with other social media analysis techniques (Simranjeet Kaur, 2023) and developing new strategies considering cultural and contextual factors. Additionally, there is a growing recognition of the need for ethical considerations in mental state inference, with researchers exploring ways to protect privacy and confidentiality. The future of mental state inference in social media holds great promise but also requires ongoing efforts to address the limitations and challenges associated with this area of research. As the field continues to evolve, researchers and practitioners must remain committed to advancing the understanding of mental state inference and its applications in mental health.

References

Abadi. (2015). TensorFlow: Large-Scale machine learning on heterogeneous distributed systems. In *Proceedings of the SIGCHI conference on human factors in computing systems* (pp. 1909–1912). ACM.

ACM Digital Library. Retrieved from ACM Digital Library. (2023).

Amanat, A., Rizwan, M., Javed, A. R., Abdelhaq, M., Alsaqour, R., Pandya, S., & Uddin, M. (2022). Deep learning for depression detection from textual data. *Electronics, 11*(5), 676. https://doi.org/10.3390/electronics11050676

Ansari, G., Garg, M., & Saxena, C. (2021). *Data augmentation for mental health classification on social media.* arXiv. https://arxiv.org.

Braghieri, L. L. (2022). *Data and code for: Social media and mental health.* American Economic Association.

Burnap, P., Colombo, G., & Scourfield, J. (2015). Machine classification and analysis of suicide-related communication on Twitter. In *HT 2015 - proceedings of the 26th ACM conference on hypertext and social media* (pp. 75–84). Association for Computing Machinery, Inc. https://doi.org/10.1145/2700171.2791023

Chancellor, S., Birnbaum, M. L., Caine, E. D., Silenzio, V. M. B., & De Choudhury, M. (2019). A taxonomy of ethical tensions in inferring mental health states from social media. In *FAT* 2019 - proceedings of the 2019 conference on fairness, accountability, and transparency* (pp. 79–88). Association for Computing Machinery, Inc. https://doi.org/10.1145/3287560.3287587

Chancellor, S., & De Choudhury, M. (2020). Methods in predictive techniques for mental health status on social media: A critical review. *Npj Digital Medicine, 3*(1). https://doi.org/10.1038/s41746-020-0233-7

Chen, Z., Yang, R., Fu, S., Zong, N., Liu, H., & Huang, M. (2023). *Detecting reddit users with depression using a hybrid neural network.* arXiv. https://doi.org/10.48550/arXiv.2302.02759

Choudhary, A. (2017). Social media data for sentiment analysis. In *Retrieved from McCormick northwestern engineering.*

Choudhury. (2021). I am predicting depression via social media. In *Proceedings of the international AAAI conference on web and social media* (pp. 128–137). PKP Publishing Services Network.

Coppersmith, G., Dredze, M., & Harman, C. (2014). Quantifying mental health signals in twitter. In *Proceedings of the annual meeting of the association for computational linguistics* (pp. 51–60). Association for Computational Linguistics (ACL). https://aclweb.org/.

Courville, I. G. (2016). *Deep learning.* MIT Press.

De Choudhury, M., Kiciman, E., Dredze, M., Coppersmith, G., & Kumar, M. (2016). Discovering shifts to suicidal ideation from mental health content in social media. In *Conference on human factors in computing systems - proceedings* (pp. 2098–2110). Association for Computing Machinery.. https://doi.org/10.1145/2858036.2858207

Deng, W., & Yang, Y. (2021). Cross-platform comparative study of public concern on social media during the COVID-19 pandemic: An empirical study based on twitter and weibo. *International Journal of Environmental Research and Public Health, 18*(12), 6487. https://doi.org/10.3390/ijerph18126487

Ferwerda, B., Schedl, M., & Tkalcic, M. (2016). *Using Instagram picture features to predict users' personality* (Vol. 9516, pp. 850–861). Springer Science and Business Media LLC. https://doi.org/10.1007/978-3-319-27671-7_71

Garg, M. (2023). Mental health analysis in social media posts: A survey. *Archives of Computational Methods in Engineering, 30*(3), 1819–1842. https://doi.org/10.1007/s11831-022-09863-z

Gupta, U., & Gupta, D. (2022). Least squares structural twin bounded support vector machine on class scatter. *Applied Intelligence*, 1–31. https://doi.org/10.1007/s10489-022-04237-1

Gupta, U., Gupta, D., & Agarwal, U. (2022). Analysis of randomization-based approaches for autism spectrum disorder. In *Lecture notes in electrical engineering* (Vol. 888, pp. 701–713). Springer Science and Business Media Deutschland GmbH. https://doi.org/10.1007/978-981-19-1520-8_57

Harrigian, K., Aguirre, C., & Dredze, M. (2021). On the state of social media data for mental health research. In *Computational linguistics and clinical psychology: Improving access, CLPsych 2021 - proceedings of the 7th workshop, in conjunction with NAACL 2021* (pp. 15–24). Association for Computational Linguistics (ACL). https://aclanthology.org/2021.clpsych-1.

Imran, A. S., Daudpota, S. M., Kastrati, Z., & Batra, R. (2020). Cross-cultural polarity and emotion detection using sentiment analysis and deep learning on COVID-19 related tweets. *IEEE Access, 8*, 181074–181090. https://doi.org/10.1109/access.2020.3027350

Katikalapudi, R., Chellappan, S., Montgomery, F., Wunsch, D., & Lutzen, K. (2012). Associating internet usage with depressive behavior among college students. *IEEE Technology and Society Magazine, 31*(4), 73–80. https://doi.org/10.1109/mts.2012.2225462

Kaur, S., Bhardwaj, R., Jain, A., Garg, M., & Saxena, C. (2022). Causal categorization of mental health posts using transformers. In *Proceedings of the 14th annual meeting of the forum for information retrieval evaluation* (pp. 43–46). https://doi.org/10.1145/3574318.3574334

Kross, E., Verduyn, P., Demiralp, E., Park, J., Lee, D. S., Lin, N., Shablack, H., Jonides, J., Ybarra, O., & Sueur, C. (2013). Facebook use predicts declines in subjective well-being in young adults. *PLoS One, 8*(8), Article e69841. https://doi.org/10.1371/journal.pone.0069841

Liu, B. (2012). Sentiment analysis and opinion mining. *Synthesis Lectures on Human Language Technologies, 5*(1), 1–167. https://doi.org/10.2200/s00416ed1v01y201204hlt016

Moreno, M. A., Jelenchick, L. A., Egan, K. G., Cox, E., Young, H., Gannon, K. E., & Becker, T. (2011). Feeling bad on Facebook: Depression disclosures by college students on a social networking site. *Depression and Anxiety, 28*(6), 447–455. https://doi.org/10.1002/da.20805

Reece, A. G., & Danforth, C. M. (2017). Instagram photos reveal predictive markers of depression. *EPJ Data Science, 6*(1). https://doi.org/10.1140/epjds/s13688-017-0110-z

Rosen, L. D., Whaling, K., Rab, S., Carrier, L. M., & Cheever, N. A. (2013). Is Facebook creating "iDisorders"? The link between clinical symptoms of psychiatric disorders and technology use, attitudes and anxiety. *Computers in Human Behavior, 29*(3), 1243–1254. https://doi.org/10.1016/j.chb.2012.11.012

Safa, R., Bayat, P., & Moghtader, L. (2022). Automatic detection of depression symptoms in twitter using multimodal analysis. *The Journal of Supercomputing, 78*(4), 4709–4744. https://doi.org/10.1007/s11227-021-04040-8

Sandoval-Almazan, R., & Valle-Cruz, D. (2020). Sentiment analysis of Facebook users reacting to political campaign posts. *Digital Government: Research and Practice, 1*(2), 1–13. https://doi.org/10.1145/3382735

Seabrook, E. M., Kern, M. L., Fulcher, B. D., & Rickard, N. S. (2018). Predicting depression from language-based emotion dynamics: Longitudinal analysis of Facebook and twitter status updates. *Journal of Medical Internet Research, 20*(5), e168. https://doi.org/10.2196/jmir.9267

Settles, B. (2009). *Active learning literature survey*. University of Wisconsin-Madison Department of Computer Sciences. CS Technical Reports http://digital.library.wisc.edu/1793/60660.

Simranjeet Kaur, R. B. (2023). Causal categorization of mental health posts using transformers. In *Proceedings of the 14th annual meeting of the forum for information retrieval evaluation* (pp. 43–46). arXiv.

Tsakalidis, A., Chim, J., Bilal, I. M., Zirikly, A., Atzil-Slonim, D., Nanni, F., Resnik, P., Gaur, M., Roy, K., Inkster, B., Leintz, J., & Liakata, M. (2022). Overview of the CLPsych 2022 shared task: Capturing moments of change in longitudinal user posts. In *CLPsych 2022 - 8th workshop on computational linguistics and clinical psychology, proceedings* (pp. 184–198). Association for Computational Linguistics (ACL).

Wendlandt, L., Mihalcea, R., Boyd, R. L., & Pennebaker, J. W. (2017). Multimodal analysis and prediction of latent user dimensions. In *Lecture notes in computer science (including subseries lecture notes in artificial intelligence and lecture notes in bioinformatics)* (Vol. 10539, pp. 323–340). Springer Verlag. https://doi.org/10.1007/978-3-319-67217-5_20

Xu, Z., Pérez-Rosas, V., & Mihalcea, R. (2020). Inferring social media users' mental health status from multimodal information. In *Lrec 2020 - 12th international conference on language resources and evaluation, conference proceedings* (pp. 6292–6299). European Language Resources Association (ELRA).

Yazdavar, A. H., Mahdavinejad, M. S., Bajaj, G., Romine, W., Sheth, A., Monadjemi, A. H., Thirunarayan, K., Meddar, J. M., Myers, A., Pathak, J., & Hitzler, P. (2020). Multimodal mental health analysis in social media. *PLoS One, 15*(4). https://doi.org/10.1371/journal.pone.0226248

Yilin, W., & Baoxin, L. (2015). Sentiment analysis for social media images. In *2015 IEEE international conference on data mining workshop (ICDMW)* (pp. 1584–1591). https://doi.org/10.1109/ICDMW.2015.142

Turcan. (2019). *Dreaddit*. Retrieved from paperswithcode: https://paperswithcode.com/dataset/dreaddit.

CHAPTER THREE

Detection of social mental disorder using convolution neural network

Mangi Lal Choudhary[1], Rajesh Prasad[2], Jayashree Prasad[2] and Nihar Ranjan[3]

[1]B.N. College of Pharmacy, B. N. University, Udaipur, Rajasthan, India
[2]School of Computing, MIT Art, Design and Technology University, Pune, Maharashtra, India
[3]Information Technology, JSPM's Rajarshi Shahu College of Engineering, Pune, Maharashtra, India

1. Introduction

Observation, interviews, and a variety of other sorts of inquiries and assessments are often employed from a biological perspective to detect mental health problems. Here are some examples of diagnostic methods: This strategy not only necessitates a greater amount of effort, but it also raises the likelihood of a late presentation as it does not aid in early identification and treatment when mental health illnesses first appear. Although using multidimensional machine learning to identify mental health concerns currently yields less accurate results than traditional mental health targeted therapies, it does have some distinct advantages. Big data has allowed the creation of an intelligent system capable of evaluating and diagnosing, using this technology, all of this can be done at the same time. Furthermore, it is possible to continually monitor its users' mental health, which is critical for recognizing potentially hazardous mood swings early on so that they may be prevented. Only artificial intelligence (AI) is required to do this. None of the concerns indicated above can be recognized by mental health examinations that are regarded as industry standards (Day et al., 2020). As a result, older, more established strategies for diagnosing mental health disorders and more current, cutting-edge techniques for detecting mental health illnesses based on massive data sets may complement one another. This is because traditional approaches have a long history of being used to identify mental health disorders. The diagnostic accuracy of big data systems falls short of traditional medical diagnosis; thus, by defining ways for obtaining both slack entry and strict exit, the model may be made more practical for use in the real world.

Emotional AI and Human-AI Interactions in Social Networking
ISBN: 978-0-443-19096-4
https://doi.org/10.1016/B978-0-443-19096-4.00012-2

Because the diagnostic accuracy of big data systems is inferior to that of traditional health diagnostics, this would be done. Therefore, if the model were used in real life, it would be more accurate. Given this, applying the model to real-world data would result in a higher degree of accuracy (Chen, 2022). This was formerly unachievable, but it is now viable due to the increased availability of data through big data platforms. This was not previously feasible. To put it another way, the big data method recognizes individuals who are supposedly suffering from psychological problems and then recommends or urges them to get professional help. In other words, the system can combine curative and preventative care. The next stage is to use conventional techniques to detect mental health disorders, after which the necessary therapies should be delivered in line with the established diagnosis. These classes will only be offered to those who have already shown that they are excellent parents. Only parents who have earlier shown very high levels of achievement in the task of managing their children will be permitted to participate in these programs (MamSharifi et al., 2021). High levels of genetic resemblance between parents almost always result in major personality disorders and other mental health issues. This raises the probability that their children may suffer from the same conditions. A mixed-methods, pragmatic approach is necessary to meet the challenges of developing and evaluating parenting programs that are more precisely tailored to the requirements of families. This is required to create and evaluate parenting programs that are customized to the requirements of individual families. There is some proof that home-schooling has a substantial influence on the mental health of teens. This result might be either beneficial or bad. Chen explored the impact of several forms of family cumulative risk on adolescent development and growth. He then developed strategies for educating families on how to prepare for such hazards. The primary purpose of Chen's research was to identify the strategies that have the most potential for assisting young people in maintaining outstanding mental health and developing into individuals of exceptional character (Mekawi et al., 2021). The purpose of the whole research endeavor was to attain this goal. They were successful in reaching this aim by using regression analysis. Whether racial discrimination causes the onset of depressive, anxious, and posttraumatic stress disorder symptoms using bivariate correlations. They were effective in determining whether racial bias contributed to the development of these symptoms. Prejudice based on a person's race has been proven in studies to have harmful consequences, even when the link between the two is questioned. (Asl et al., 2022) investigated how critical thinking may be utilized to predict occupational self-efficacy. This was accomplished using regression analysis. They concluded that one's ability to think critically has a significant impact on one's degree of self-efficacy at work. It is recommended that a person with mental health consult with a mental health professional to determine whether they are depressed. This strategy has been employed for a long time. These interviews, in addition to being tiresome and time-consuming, represent a major danger to the health of individuals who choose to participate in them. Given the facts stated above, there is little reason to be optimistic about the possibility of avoiding or curing depression. In the next few years, there will most certainly be a higher need for computer algorithms that can analyze a person's mental state using a large amount of data. We began by defining and computing emotional dynamic characteristics (Gong et al., 2018), which enabled us to generate more exact predictions about the components that determine how emotional states are created. This was done to help us understand how the brain generates different emotional states. This was done to have a better grasp of how the brain functions. We did this to get a better understanding of how distinct emotional states arise throughout time. More specifically, the researchers sought to discover whether this issue was in any way connected to the topic. They were able to predict the likelihood that gamers would exhibit personality traits, therefore. They wanted to investigate whether there was a possibility that video game players might exhibit personality characteristics (Adamopoulos et al., 2018). The findings of deep learning—trained deep neural networks demonstrated a much better probability of properly forecasting a player's attribute level than just the chance of accomplishing this (Ammannato & Chiesi, 2020). This was in comparison to the chance of correctly recognizing a player's feature level. This was applicable to five of the six personality development elements studied and published a groundbreaking, deep learning-based solution for tracking a child's stress level while exercising by monitoring their body temperature, movement speed, and the amount of sweat they produced (Nair & Bhagat, 2020). This gadget can measure a person's stress level by detecting their core temperature, movement speed, and sweat generation. This device can determine how stressed a person is based on their core temperature, the velocity of activity, and the quantity of perspiration produced. This action was taken to attain the aim of establishing realistic coping strategies for dealing with mental stress in the context of revolutionary medical approaches, with the end goal of expanding the number of individuals who could really benefit from these methods. The activity was carried out to achieve the objective of establishing effective coping methods for dealing with mental stress in the context of reducing medical treatment. Deep neural networks may construct mental representations based on the mechanism of the network's real-world inputs (Kashyap, 2019). These representations might potentially be utilized to store mental pictures in the future. However, psychological research focuses on how people develop specific traits and self-categories. Peterson and his colleagues remarked that convex optimum (Rachakonda et al., 2019) may give easy tweaks that may account for these variations to widen the area of psychological inquiry and computer modeling. The context for this finding includes the advancement of the area of psychology study and the utilization of computer modeling. Despite the possibility that there is a considerable association between mental health difficulties and activity in the study that was addressed, the labels obtained via the use of surveys tend to be incorrect. This is the case despite the reality that there is a considerable association between the two. Despite the reality that the two are good friends, this situation has arisen. Furthermore, since psychological well-being is linked to a particular source of behavioral data, conducting an appropriate evaluation of the mental health state is difficult (Sameh et al., 2019). This is because it is difficult to determine whether mental health is to blame for the observable traits, this is because mental health may be altered by a broad variety of factors.

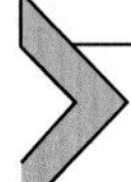

2. The application of deep learning in personality analysis and work-based learning

This piece of psychological research is being combined with advances in very sophisticated AI. Many companies that specialize in this field are already developing products that combine enhanced and virtual reality technologies. Those receiving mental health therapy using these tools may have the opportunity to see video images in virtual or augmented reality as part of their treatment. Because of the growth of emerging technologies such as the Internet, big data, and AI, psychologists now have access to new resources. Furthermore, this has opened exciting new study avenues. Gadgets such as smartphones, social media sites, sensors, and other electronic equipment provide access to a lot of data about people's daily activities. Using this information, users may get personalized content and advertisements. Fig. 3.1 highlights and organizes the huge quantities of data that mental health has gathered.

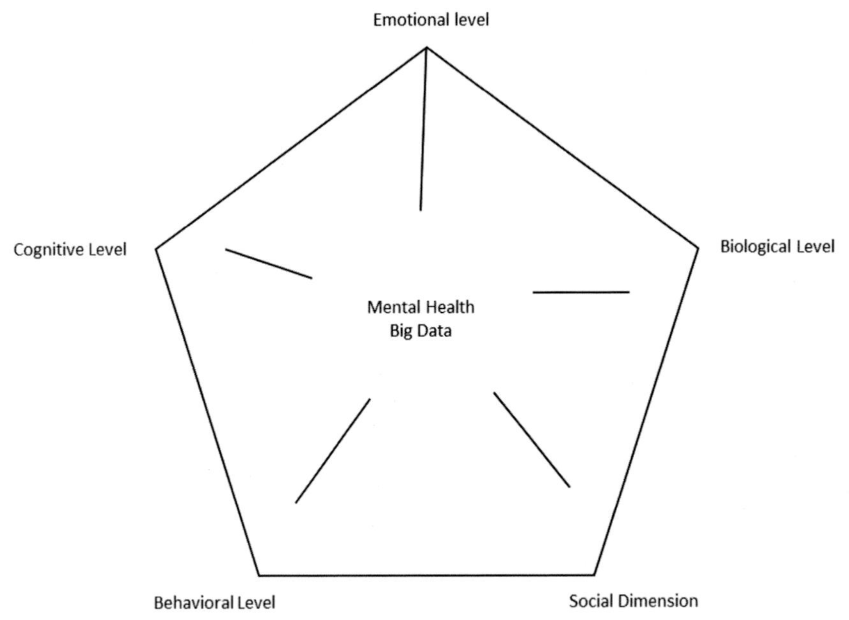

Figure 3.1 The massive datasets collected at the mental health level as separated into categories: Emotion, memory, behavior, and society.

Using these big datasets, it is possible to create predictions about people at the mental health level. They do this to convey their feelings to their friends and to let people know what they are feeling. A presentation demonstrating the various ways in which people can express themselves to one another (Peterson et al., 2018). Wearables, like sleep bracelets, might be used to gather data on women's sleeping habits, which could subsequently be utilized in research to get insight into mental health concerns. Using these tools may help individuals get a better understanding of their own sleeping patterns. In other words, if the mental health statistics are of sufficient quality, the analysis of these massive volumes of mental health data may reveal individuals' emotions and social circumstances.

The five most important parts of a person's past, as indicated in Fig. 3.2, are critical to describing that person's personality. This category encompasses a wide range of personality qualities, including psychoticism, openness, accountability, openness, and life satisfaction (Ge et al., 2021). This is true even though AI cannot yet totally replace human psychiatric counselors. The implementation of AI in business and marketing for the mental health sector has several prospects for growth and improvement. AI will 1 day

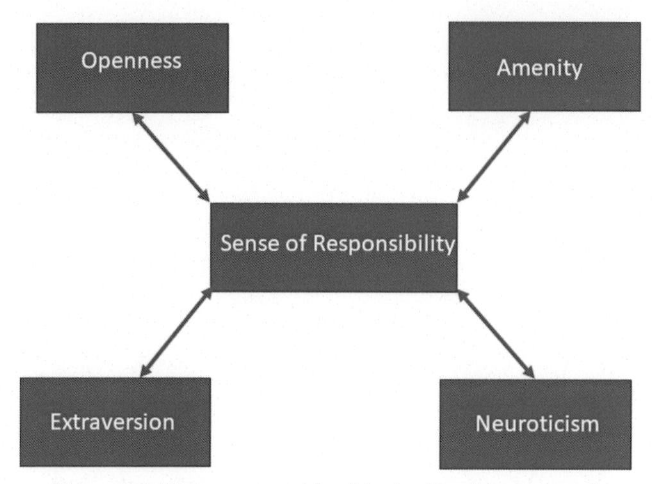

Figure 3.2 Human mental health classification approach.

replace expensive and inefficient mental health therapies, so this technology will help AI progress (Mostowik et al., 2021). AI will also replace some expensive and unsuccessful mental treatments.

High-dimensional datasets may include a substantial number of distinct features. The purpose of clustering algorithm research is to determine which attributes among a vast number of characteristics should be separated into distinct groups first (Shemeis et al., 2021), and this determination is done via the study of clustering techniques. One method is to select which items in a vast collection should indeed be sorted into groups first. The problem that must be addressed is determining how to pick the most useful attribute. Formulas are one of the numerous approaches for calculating the entropy of information. The more sophisticated the database, the less valuable it is. Entropy (E) is a measure of the likelihood of something occurring. The letter I, which stands for "total," stands for all the classes represented by the letter E. The method provided here may be used to calculate the gain in knowledge attributable to set E. The formula looks like this: It is feasible to get the most advantage by selecting the appropriate attributes. To give you only one example: When the two methods are examined side by side, it is evident that the Classification and Regression Trees (CART) approach is substantially more effective than the previous one. Because of the way they are constructed, decision trees are independent of the information that is provided. It is possible to process unrelated attributes alongside nominal and numerical data (Schweizer, 2020).

Therefore, creating content that is true while also taking other points of view into account is considerably more difficult. Some of the drawbacks connected with decision trees may be solved by using a technique known as a xgboost tree. Horizontal stripe trees are used to mitigate the detrimental impact of a growing number of weak classifiers on tree structures (Ahmadi et al., 2020). This may be accomplished using several classifiers, each having a different scale of boost. It shows one way to find the negative slope, which is then used as the residual value for each of the seven basic classifiers. There are many ways to write out the support vector tree.

Every tree in a randomized forest is built without regard for the consequences of previous seedlings. A gradient-boosted tree, on the other hand, is produced while taking into consideration the results of the forests that came before it. A random forest, on the other hand, builds each distinct plant without considering the results of the plants that came before it (Bako, 2020). A random forest, on the other hand, is one in which each tree is formed independently, without regard for the consequences of the branches that came before them in the chain. The use of a wide range of linear models may be useful for both the creation of complex features and the modification of features. Using the second strategy, it is feasible to ensure that the feature will be retained. Support vector trees, which are remarkably equivalent to randomized forests, could be used to alleviate the overfitting and lack of compassion for outliers that decision trees have. This issue arises because selection trees store information in a sequential fashion, making eventual access to the data more difficult.

Using the approach provided in Example, you can calculate the error produced by the simple classifier by applying it to the distribution. This will enable you to assess the accuracy of the basic classifier. The technique may be used to calculate the contribution of the base classifier to the overall performance of the final classifier. This may be performed by comparing the early classifier results to the final classifier results (Zhao et al., 2018). Then, the procedure should continue until either your classification effect's accuracy reaches the necessary level or you have utilized every feature that the provided basic classifier has to offer. There is also the information accumulated by the individual responsible for the initial categorization, such as utilizing a person's name and image to identify any and almost every computer that goes to them. Fig. 3.3 depicts the general structure of the data-driven, multisource technique for uncovering student concerns regarding mental health difficulties, which can be found further on this page.

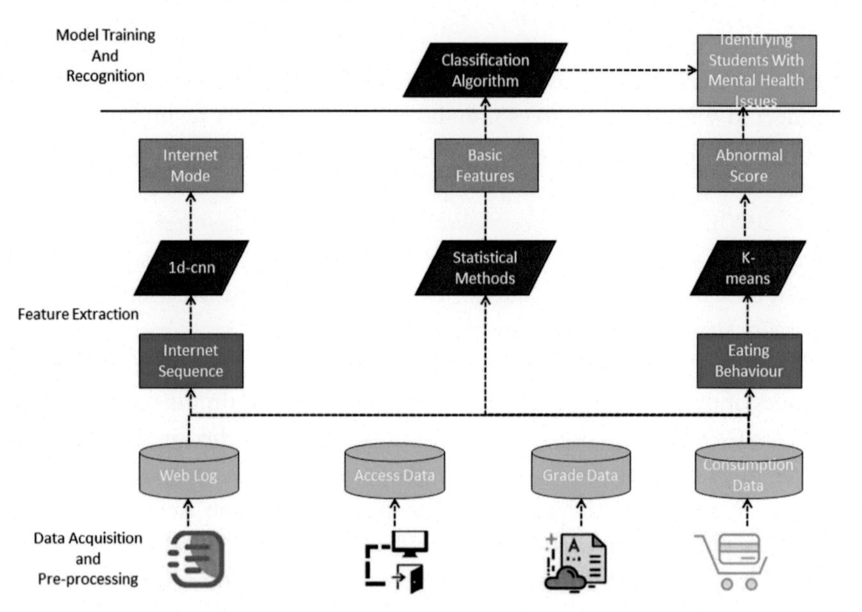

Figure 3.3 The proposed method for identification of mental health issues.

Even the highest recall rate of 0.58 indicates that only a very small percentage of the samples were correctly identified, as demonstrated. Whenever the two rates are compared, this can be observed (Zhao et al., 2018). The recall rate was also higher than the precision and recall rates. First, the logistic regression outperformed the others in terms of overall performance, notably in terms of accuracy rate (recall), which rose after earlier changes. As a result, the mental health identification strategy, which collects data from several sources, has chosen the decision tree as its principal method of classification. According to the study done on the test set, the algorithm that integrated data from several sources into a single, coherent whole properly recognized 56% of the children as having mental health difficulties. The results of the experiments done on the test set were used to find out this information.

2.1 An algorithm assessing the mental health of college students

This recognition system evaluates a variety of characteristics, including the student's grade point averages, the amount of material they consume, the degree of internet activity, and entry constraints, to draw conclusions about their classroom behavior. These data may subsequently be used to infer the students' behavior. This study increased the algorithm's capacity to discern between the many types of mental illness that could afflict young people. Unfortunately, further investigation showed that the proposal included several problems that needed to be solved before it was implemented (Lokhandwala & Spencer, 2022). Even though the multisensory data-based identification algorithm may be used to differentiate youngsters with mental health concerns, the experiment's findings are not as powerful as they may be. While it is possible to identify students who have mental health concerns, the experiment does not provide the greatest outcomes. After doing a more complete investigation of the approach architecture, two flaws were discovered (Abdullah et al., 2021). Because the time necessary to complete offline behavior sequences is substantially greater than the time required to complete online behavior sequences, there is an abundance of potential starting points from which to pick. It is quite unlikely that a series shortened to a shorter period can teach us anything about youngsters' internet behavior. Second, the online course has been taught twice, each time resulting in a drop in the amount of available income. When compared to training with a classification approach, utilizing an ID-CNN to extract surfing patterns results in one loss (Ramirez-Asis et al., 2022). This modification is the consequence of more successful surfing pattern extraction using an ID-CNN. These expenditures and losses accumulate, while the therapy is being delivered. To achieve this aim, we will first create an online trajectory matrix, and then we will create a network to better our ability to spot mental health concerns that risk children. These two activities together will bring us one step closer to our objective (Barnicot, 2021).

2.2 Build a student online trajectory matrix for your class

The network log includes time stamps that are accurate to the second level. If you keep producing the same sequence in the same order over time, you will eventually wind up with sequences of varying lengths. This might have unanticipated consequences. Students often participate in time-sensitive work while online, and they may provide various excuses for doing so during class. The online activities in which students participate usually have a time constraint. Students, for example, are more likely to visit academic websites during the day but more likely to visit recreational and retail websites at night (Abdullah et al., 2021). If we categorize the students' Internet usage by time of day, we may be able to get a clearer sense of the conditions under which they were using the internet for a variety of purposes. The day was split into 24 different segments, each lasting precisely 1 h. This approach converts a child's typical internet surfing behavior into a time-and-activity matrix (web browsing category). If the search for access records within the specified time fails, the value 0 will be entered into the field. The most common method for extracting features from matrices with only two dimensions is to use a convolutional neural network with only two dimensions. This is because of the ease with which such networks may be trained. Fig. 3.4 depicts the whole building in its entirety.

The model's purpose is to help students get more out of their online time, the second module constructs a totally linked neural network by using the main attributes of the four different data sources as input (including anomalous scores). It is possible to generate an impact that baseline student attributes have on recognizing mental health illnesses using this network. After the outputs of the modules have been merged, an artificial neural network made up of layers that are connected at every point is employed. Each variable in the model has some form of link with every other variable, either directly or indirectly.

2.3 Network architecture design

The live behavior trajectory module's two-dimensional convolutional neural network is developed using eight distinct neural network layers. It is recommended that the edges of the matrix be filled in with 0s to guarantee that the dimensions of the matrix are kept after the convolution method. A batch

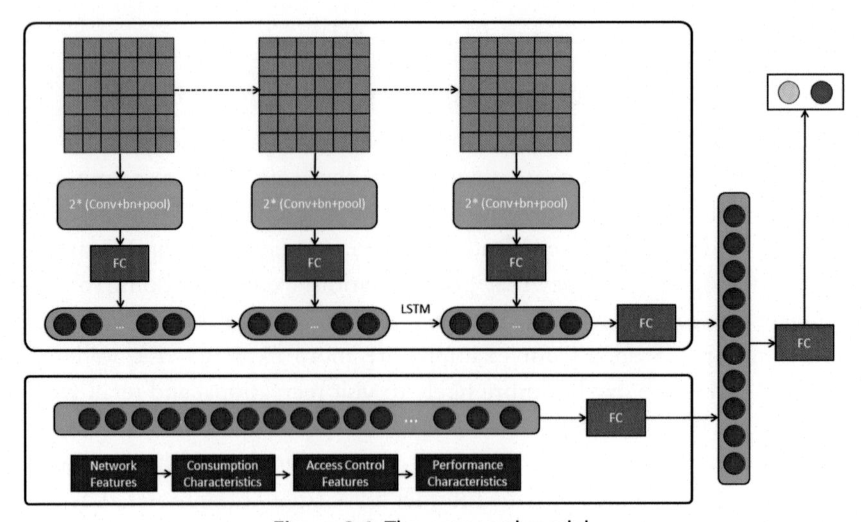

Figure 3.4 The proposed model.

normalization layer is employed in the neural network's second and fifth layers. Before proceeding to the fully connected layer, the recovered feature matrix is vectorized to ensure that it may be utilized later in the analysis. When training the model, it is critical to avoid overspecializing it as much as possible since doing so considerably increases the model's ability to generalize the results of its analysis. As a result, the eighth level employs a dropout network layer to deal with this situation. Each pupil is represented by their own separate vector in this illustration. Several aspects of each student's daily surfing activity are gathered and examined as soon as the 2D-CNN phase is completed (Nair, Alhudhaif, et al., 2021). The ninth hidden layer comprises the calculation required to identify the temporal dependence. In a computer network containing both long-term and short-term memory, there is always some form of output being processed simultaneously.

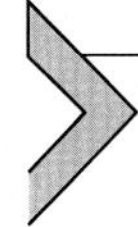

3. Experiments employ personality drafts and data

3.1 Social network data triangulation for trait prediction and profiling

The five personality prediction models were created using the conventional support vector machines (SVM) methodology. This step follows the removal of features that did not fit the ideal model using SVM during the training process. The scikit-learn package, which was also employed throughout the experiment, was able to contribute to the development of the feature selection process. During training, grid search, and tenfold cross-validation are employed to determine the optimal hyperparameter values for the linear SVM (Mohanakurup et al., 2022). This is done to ensure that the model is as accurate as possible. The system's RFE interface is made available to make this aim simpler to achieve. Participants in this study are asked to classify people using a list of 42 unique visual traits. In other words, 42 models were created during the algorithm's initial training iteration, and one feature was removed from the models that remained at the end of each training iteration. All 42 models were trained using the default settings, which included grid search for the optimal hyperparameters and ten-fold cross-validation. This operation has been completed on all the models (Parashar et al., 2022). With this technique, the algorithm can construct, at each iteration of its operation, the model that is the most accurate representation of the data that it is now capable of supplying, given the information now at its disposal. When all SVM-RFE iterations are complete, the most accurate model will be constructed, and features will be chosen depending on how well they match the model. The next step is to run an experiment using the data that were collected. Up until this point, the only variable that has changed is the phase that involves feature extraction; all other parts have stayed constant.

3.2 The findings of an experiment regarding openness and extraversion

It is quite a marvel that the deep convolutional neural network used to obtain 18 of the 20 picture-style features could tie back to the traits discussed in the introduction. The researchers concluded that an avatar's aesthetics, rather than its face, has a substantial impact on the openness model. In terms of accuracy and F1 score, the model developed using style attributes outperformed the model built with facial characteristics. An unbiased third party provided feedback on the individual's facial traits as well as their appearance preferences. These judgments are confirmed by an examination of the individual's facial characteristics and sense of fashion. Figs. 3.5—3.6 show the

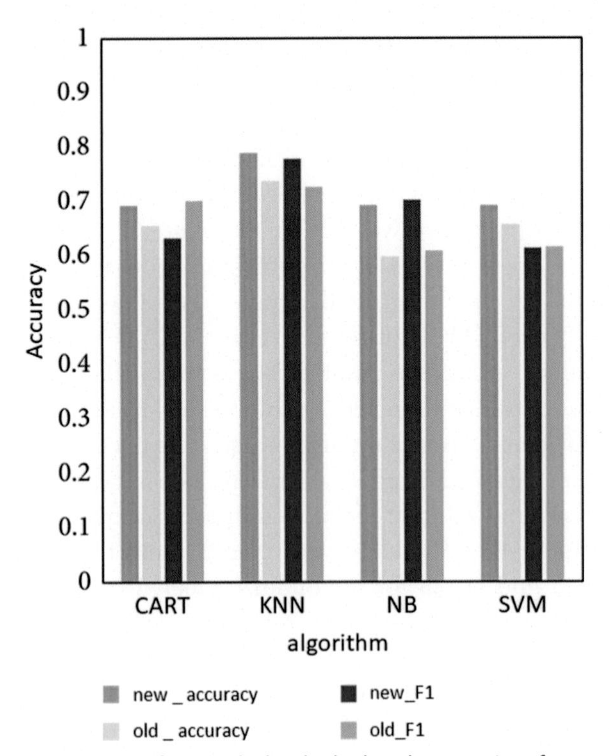

Figure 3.5 The outcomes of research that looked at the capacity of openness and extraversion to make correct predictions.

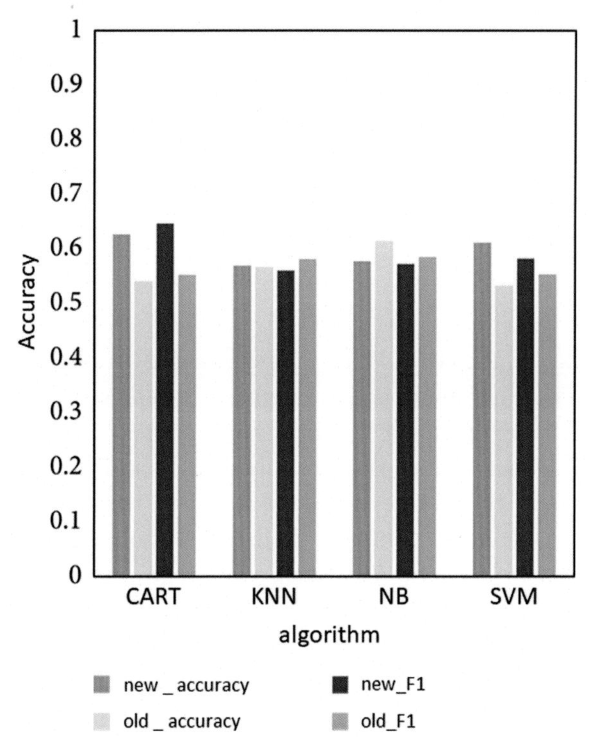

Figure 3.6 The outcomes of research that looked at the capacity of extraversion to make correct predictions.

outcomes of research that looked at the capacity of openness and extraversion to make correct predictions.

The accuracy rate of the four open projections increased from 66.0% to 70.6% because of this modification. An increase in average accuracy could explain both data points. The predicted extroverts went up from 56.3% to 59.6%, the F1 seemed to be more accurate, going from 56.8% to 59.0%, and model performance got better. Additionally, albeit to a lesser degree, the accuracy of the SVM classifier with a Gaussian kernel has increased, and this development should be brought to your attention (Nair, Vishwakarma, Soni, Patel, & Joshi, 2021). The accuracy of the CART classification model, on the other hand, has hardly altered, while the F1 value has gotten less accurate. In contrast, we have the F2 value case here. The classification of an individual's prediction of openness in personality achieved an accuracy of 78.4% as well as an F1 value of 77.8% when SVM-RFE was used as an extra feature selection approach. A variety of methodologies are used to

Table 3.1 Summarizes the outcomes of an online survey on different levels of trust.

Classification algorithm	Image set 1		Image set 2	
	Accuracy	F2	Accuracy	F2
Support vector machine	79.4	72.4	93.7	89.9
NB	79.3	80.3	86.1	84.4
k-Nearest neighbor	89.5	88.9	94.4	93.9
Classification and regression trees	91.5	88.4	89.3	91.8

analyze the experimental outcomes of openness prediction for a broad range of user-provided pictures. Images are supplied by a diverse group of people.

Table 3.1 shows that using the classification model to infer user character from multiple photographs uploaded daily by users and using it to infer user openness from various pictures uploaded daily by users improved the classifying model's overall performance. Both applications have been demonstrated to improve the model's overall effectiveness. Both uses have been found to improve a woman's overall performance. Table 3.2 demonstrates that when a classifier was used to estimate user openness from the large number of images uploaded by users on a regular basis, their actual quality improved. This is evidenced by the categorization model's superior overall performance. This is shown by the fact that doing so resulted in a classifier that scored better in general. The F1 models' average accuracy climbed from 68.1% to 74.5, the only component of the CART classification model to show a decrease was the F1 value, and even that drop was just 0.05, making it virtually inconsequential. All other parameters remained unchanged. This procedure produced the best potential results; hence, it was selected. These are the highest possible scores that a player has a chance of attaining. F1 and the accuracy rate both increased during the same time, with F1 increasing by 0.176 and the accuracy rate increasing by 0.133, respectively. These two augmentations occurred concurrently. This revealed that predicting a user's

Table 3.2 Summarizes the extraversion study's results accessible online.

Classification algorithm	Image set 1		Image set 2	
	Accuracy	F1	Accuracy	F1
Support vector machine	72.4	69.1	82.1	87.2
NB	71.3	68.8	82.4	81.9
k-Nearest neighbor	78.4	71.5	83.4	80.7
Classification and regression trees	90.2	87.9	91.3	89.8

openness features based on photographs uploaded on a regular basis is both doable and effective. It was also shown that when numerous pictures are used to forecast a user's openness, the model's performance improves. This was shown by the model's performance when many photographs were used to anticipate a user's openness. This was a two-fold example since it demonstrated the possibility of merging data from numerous images into a single forecast as well as the probability that the model would perform better when multiple shots were used to estimate a user's openness. This was shown by the fact that the model performed better when many photographs were used to predict how open a user would be. This demonstrated that the hypothesis is valid.

Even though the average value of F1 increased over the course of the inquiry from 59.0% to 71.0% shown in Table 3.2. This is shown by the fact that the model's overall performance increased. Every single categorization model is more than 70% accurate. This is due to improvements in each of these indicators, which explains why this is the case. The two evaluation criteria that are being used may have the most visible influence on overall development.

3.3 Conscientiousness test results

Throughout the duration of the SVM-RFE research project, 19 picture characteristics were disregarded. Fig. 3.7 shows the effect of deleting each of these image features on the accuracy.

After everything was said and done, just eight facial features remained. The support vector machine representation of faces SVM-RFE method was used to pull out 11 image-style descriptors and the five facial features that went with each one. When the results of several trials were compared to the results of the four improved conscientious prediction models, the average accuracy went up from 63.50% to 65.3%. Both models were employed throughout the statistical categorization process. The F1 score for conscientiousness went up to 70.2% because the way qualities were found was changed. Furthermore, the maximum predicted accuracy rose to 69.2%, when the daily image predictions from users are used, the same model has the same effect on both kinds of images. This is because, as the quantity of images increases, more data can now be used to train the models. This is due to the increased accessibility of photos. In the conscientious prediction test, the k-nearest neighbor model continues to do the worst of all categorization models, which is the same as what happened in previous experiments with avatars.

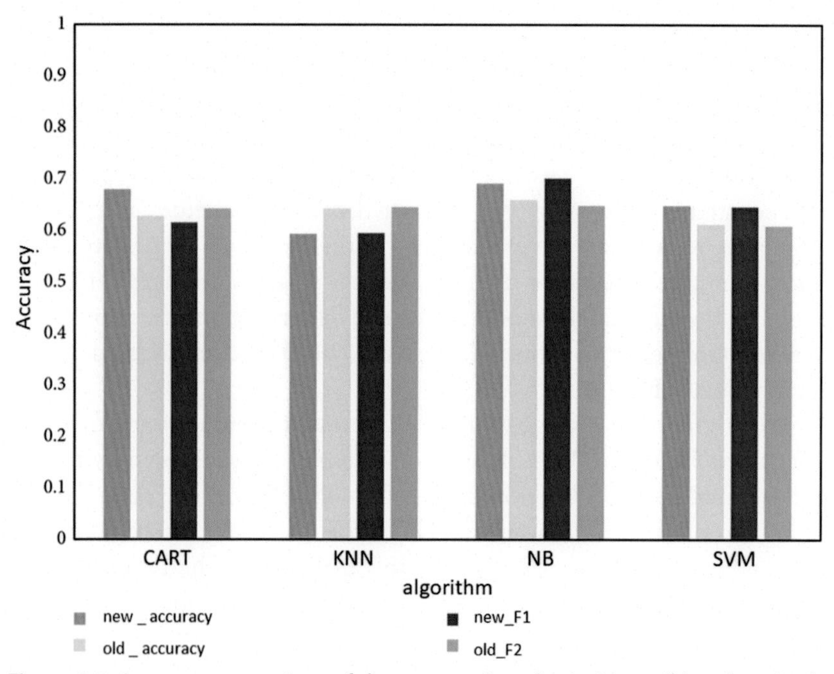

Figure 3.7 Accuracy comparison of the proposed method with traditional methods.

The Gaussian kernel SVM achieves the greatest degree of accuracy gain, 0.134. The CART classification algorithm made the F1 value go up by 0.139. This was the most noticeable change. The fact that we have reached this point demonstrates significant progress in our job. The results of the experiment show that classification models are very good at predicting conscientiousness when they are used to look at many user-submitted photos. The experimental approach and visual features were demonstrated to be useful in real-world scenarios. Even though the characteristics used in the experiment were helpful, the fact that the most accurate prediction was only 80.0% of the time shows that these characteristics have missing features.

3.4 Recognizing the nature and scope of students' mental health obstacles

DeepPsy is a model that shows how to use network parameters by going through the levels of the network in the right order. For model training optimization, the Adam optimization strategy is used, with 60 iterations and a batch size of 4. The experimental data and model results produced

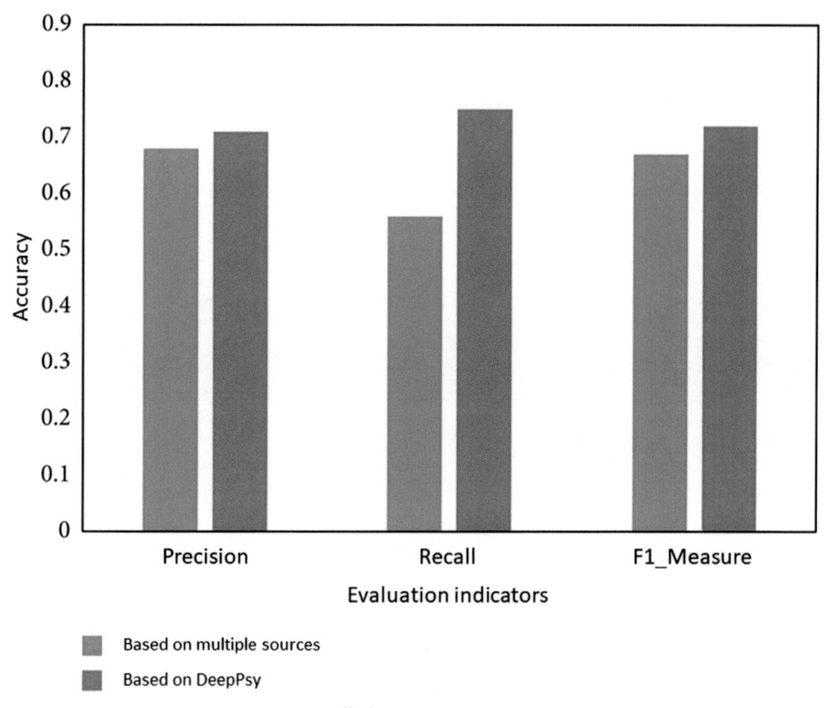

Figure 3.8 All three assessment metrics.

by DeepPsy were compared. We were able to compare the intuitive understandings of the different groups in a simpler way by providing the data in a graphical format. Fig. 3.8 shows the results of all three metrics.

After careful study, the look of the symbol used to show the recall function has been improved. The goal is to find out which children have been affected by mental health so those who need help can get it. The ability of the model to be remembered was emphasized much more than in the past. DeepPsy generated a recall of 0.75, which is 19 percentage points higher than the multisource approach. Seventy-five percent of the children may be recognized as having mental health requirements that are not being met by the approach. It is unknown what impact each of these many variables had on the outcomes. The feature set This includes network capabilities, feature tracking, feature reporting, and access and consumption monitoring. Use, speed, and access control were all parts of the network in feature set I that was taken out of feature set II. None of them will appear in any way in Feature Set II. Based on the traits shown in set II, set IV is made up of only access restrictions and has no features related to the act

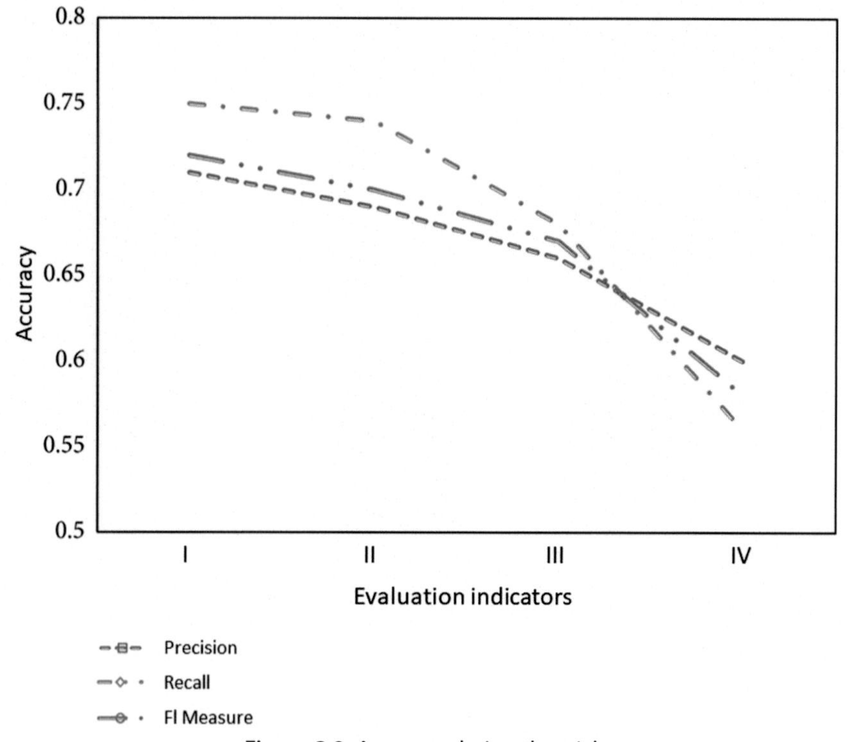

Figure 3.9 Accuracy during the trial.

of eating itself. Fig. 3.8 depicts the visual results of an experiment comparing four groups of qualities.

This is because there are so many parts that need to be used up, and each one is important to the final output. The batch size parameter was set to n = 1, 2, 3, 4, and 5 for each of the 60 iterations. Fig. 3.9 illustrates the accuracy. If the other variables are not changed, this is the situation.

This is true while all other factors remain constant. Fig. 3.10 depicts the performance of the model on experimental data.

4. Conclusions

Most of the time, the problems homeless people face are not as bad or as widespread as what national surveys show. In recent years, web server logs have become the most important way to figure out what is wrong with mental health because students engage in so many other activities besides their online behavior, this approach does not adequately capture the

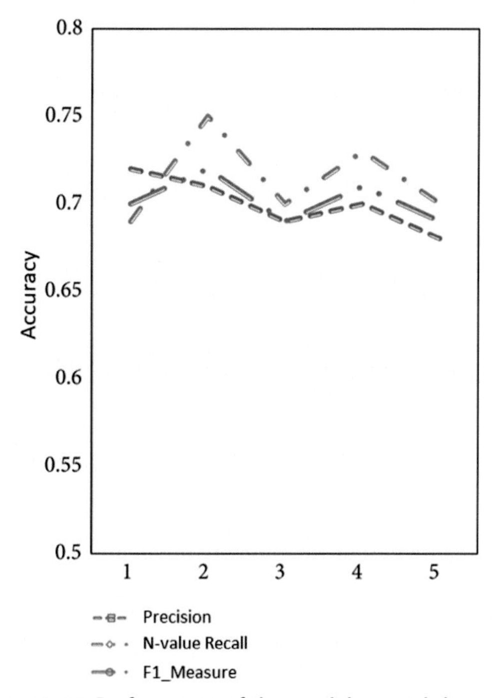

Figure 3.10 Performance of the model on trial datasets.

spectrum of psychological activity that happens while they are on campus, even though it eliminates the flaws of questionnaire surveys. When adopting the Internet-based approach to diagnosing mental health diseases, it is still necessary to complete questionnaires to get labels. This means that the labels cannot be trusted. This article talks about a mental health strategy for finding problems that use the DeepPsy model and data from several sources to find widespread problems that affect the mental health of college students. The next step in character data analysis is to use user avatars to build a deep-learning network. This will be accomplished through the application of deep learning. This system must connect key operations to patterns of online activity. The results of the test showed that the proposed strategy worked better in real-world situations than other options. In this study, the strategy used to find children who didn't know about their mental health problems and character flaws worked well. So, it is possible for young children to get psychiatric help right away so that their condition does not get worse. Researchers think that if they learn more about the subject, they will have a better idea of how much the problem affects the mental health of students.

References

Abdullah, A., Muhammad Amin, H., Abdurrahman, F., Idris, J., & Marthoenis, M. (2021). Physiological predictors of mental disorders among police officers in Indonesia. *Asia-Pacific Journal of Public Health, 33*(8), 888–898. https://doi.org/10.1177/10105395211027752

Adamopoulos, P., Ghose, A., & Todri, V. (2018). The impact of user personality traits on word of mouth: Text-mining social media platforms. *Information Systems Research, 29*(3), 612–640. https://doi.org/10.1287/isre.2017.0768

Ahmadi, M., Faramarzi, M., Basirat, Z., Kheirkhah, F., Chehrazi, M., & Ashabi, F. (2020). Mental and personality disorders in infertile women with polycystic ovary: A case-control study. *African Health Sciences, 20*(3), 1241–1249. https://doi.org/10.4314/ahs.v20i3.28

Ammannato, G., & Chiesi, F. (2020). Playing with networks. *European Journal of Psychological Assessment, 36*(6), 973–980. https://doi.org/10.1027/1015-5759/a000608

Asl, N., Dokaneifard, F., & Jahangir, P. (2022). Presenting a model for predicting emotional divorce based on personality traits and job self-efficacy mediated by perceived social support in working married women. *Journal of Counseling Research.* https://doi.org/10.18502/qjcr.v20i80.8844

Bako, A. (2020). Teaching conceptions of inquiry-based learning among Nigerian secondary school physics teachers. *International Journal of Psychosocial Rehabilitation, 24*(5), 5475–5482. https://doi.org/10.37200/ijpr/v24i5/pr2020254

Barnicot, K. (2021). Evidence-based psychological interventions for borderline personality disorder in the United Kingdom—who falls through the gaps? *Borderline Personality Disorder, 5*(2), 148–167. https://doi.org/10.33212/jpt.v5n2.2020.148

Chen, Y. (2022). An analysis of the influence of family cumulative risks on adolescents' mental health. *Advanced Journal of Nursing, 2*(3), 52. https://doi.org/10.32629/ajn.v2i3.607

Day, C., Briskman, J., Crawford, M. J., Foote, L., Harris, L., Boadu, J., McCrone, P., McMurran, M., Michelson, D., Moran, P., Mosse, L., Scott, S., Stahl, D., Ramchandani, P., & Weaver, T. (2020). An intervention for parents with severe personality difficulties whose children have mental health problems: A feasibility RCT. *Health Technology Assessment, 24*(14), 1–188. https://doi.org/10.3310/hta24140

Ge, X., Li, X., & hou, Y. (2021). Confucian ideal personality traits (Junzi personality): Exploration of psychological measurement. *Acta Psychology Sinica, 53*(12), 1321. https://doi.org/10.3724/sp.j.1041.2021.01321

Gong, Y., Lu, N., & Zhang, J. (2018). Application of deep learning fusion algorithm in natural language processing in emotional semantic analysis. *Concurrency and Computation: Practice and Experience, 31*(10). https://doi.org/10.1002/cpe.4779

Kashyap, R. (2019). Security, reliability, and performance assessment for healthcare biometrics. *Advances in Medical Technologies and Clinical Practice*, 29–54. https://doi.org/10.4018/978-1-5225-7525-2.ch002

Lokhandwala, S., & Spencer, R. M. (2022). Relations between sleep patterns early in life and brain development: A review. *Developmental Cognitive Neuroscience, 56*, Article 101130. https://doi.org/10.1016/j.dcn.2022.101130

MamSharifi, P., Sohrabi, F., Seidi, P., Borjali, A., Hoseininezhad, N., Rezaeifar, H., & Asadi, N. (2021). *The role of making voluntary function in prognosticating and preventing from addiction potential: A survey study among some members of the Iranian red crescent society.* https://doi.org/10.21203/rs.3.rs-143002/v1

Mekawi, Y., Hyatt, C. S., Maples-Keller, J., Carter, S., Michopoulos, V., & Powers, A. (2021). Racial discrimination predicts mental health outcomes beyond the role of personality traits in a community sample of African Americans. *Clinical Psychological Science, 9*(2), 183–196. https://doi.org/10.1177/2167702620957318

Mohanakurup, V., Parambil Gangadharan, S. M., Goel, P., Verma, D., Alshehri, S., Kashyap, R., & Malakhil, B. (2022). Breast cancer detection on histopathological images using a composite dilated backbone network. *Computational Intelligence and Neuroscience, 2022*, 1–10. https://doi.org/10.1155/2022/8517706

Mostowik, J., Rutkowski, K., Ostrowski, T., & Mielimąka, M. (2021). Time perspective among patients diagnosed with neurotic and personality disorders – an exploratory, clinical study. *Timing and Time Perception, 9*(3), 315–334. https://doi.org/10.1163/22134468-bja10029

Nair, R., Alhudhaif, A., Koundal, D., Doewes, R. I., & Sharma, P. (2021). Deep learning-based covid-19 detection system using pulmonary CT scans. *Turkish Journal of Electrical Engineering and Computer Sciences, 29*(SI-1), 2716–2727. https://doi.org/10.3906/elk-2105-243

Nair, R., & Bhagat, A. (2020). Healthcare information exchange through blockchain-based approaches. *Transforming Businesses With Bitcoin Mining and Blockchain Applications*, 234–246. https://doi.org/10.4018/978-1-7998-0186-3.ch014

Nair, R., Vishwakarma, S., Soni, M., Patel, T., & Joshi, S. (2021b). Detection of covid-19 cases through X-ray images using hybrid deep neural network. *World Journal of Engineering, 19*(1), 33–39. https://doi.org/10.1108/wje-10-2020-0529

Parashar, V., Kashyap, R., Rizwan, A., Karras, D. A., Altamirano, G. C., Dixit, E., & Ahmadi, F. (2022). Aggregation-based dynamic channel bonding to maximise the performance of wireless local area networks (WLAN). *Wireless Communications and Mobile Computing, 2022*, 1–11. https://doi.org/10.1155/2022/4464447

Peterson, J. C., Abbott, J. T., & Griffiths, T. L. (2018). Evaluating (and improving) the correspondence between deep neural networks and human representations. *Cognitive Science, 42*(8), 2648–2669. https://doi.org/10.1111/cogs.12670

Rachakonda, L., Mohanty, S. P., Kougianos, E., & Sundaravadivel, P. (2019). Stress-lysis: A DNN-integrated edge device for stress level detection in the IOMT. *IEEE Transactions on Consumer Electronics, 65*(4), 474–483. https://doi.org/10.1109/tce.2019.2940472

Ramirez-Asis, E., Bolivar, R. P., Gonzales, L. A., Chaudhury, S., Kashyap, R., Alsanie, W. F., & Viju, G. K. (2022). A lightweight hybrid dilated ghost model-based approach for the prognosis of breast cancer. *Computational Intelligence and Neuroscience, 2022*, 1–10. https://doi.org/10.1155/2022/9325452

Sameh, D., Khoriba, G., & Haggag, M. (2019). Behaviour analysis voting model using social media data. *International Journal of Intelligent Engineering and Systems, 12*(2), 211–221. https://doi.org/10.22266/ijies2019.0430.21

Schweizer, P. (2020). Thinking on thinking: The elementary forms of mental life neutrosophical representation as enabling cognitive heuristics. *International Journal of Neutrosophic Science*, 63–71. https://doi.org/10.54216/ijns.020201

Shemeis, M., Asad, T., & Attia, S. (2021). The effect of big five factors of personality on compulsive buying: The mediating role of consumer negative emotions. *American Journal of Business and Operations Research*, 5–23. https://doi.org/10.54216/ajbor.020102

Zhao, N., Cao, Y., & Lau, R. W. (2018). What characterizes personalities of graphic designs? *ACM Transactions on Graphics, 37*(4), 1–15. https://doi.org/10.1145/3197517.3201355

CHAPTER FOUR

Social Neuroscience: AI for education

Manvendra Singh
Department of Government and Public Administration, Lovely Professional University, Phagwara, Punjab, India

1. Introduction

The field of Social Neuroscience (SN) is still in its infancy, but it has made substantial development during the last three decades. This expanding field of interdisciplinary research, which tries to understand the biological components underlying social structures, processes, and behavior, has found applications in a variety of social settings, including education, health, and public policy. The field of cognitive neuroscience shifted its focus to this new discipline, and it was this move that eventually led to the formation of a new paradigm in the field known as Social Cognitive Neuroscience (Cacioppo & Decety, 2011). This method attempts to analyze phenomena by considering their interactions at the levels of society, cognition, and neurological system. Because of its emphasis on the importance of environmental elements in understanding human behavior, SN has attempted to oppose the primarily biological perspective held by certain schools of thought by including its theory of multilevel analysis (Cacioppo et al., 2011). This is because SN places a high value on the subject. These varied endeavors all have the same goal, which is to get a better understanding of human behavior. Many scientific articles have been written because of SN research, the vast majority of which have been published in the field of neuroendocrinology. Oxytocin and arginine vasopressin have a wide range of effects on behavior, including attachment, social recognition, and aggressive behavior.

As a result of the SN discoveries, researchers have been looking for human behaviors that are like animal behaviors (Carré & Olmstead, 2015). It is worth noting that, contrary to popular belief, oxytocin does not always enhance trust and prosocial behavior in people. Depending on the person's diagnosis as well as the presence of persistent interpersonal insecurities and

Emotional AI and Human-AI Interactions in Social Networking
ISBN: 978-0-443-19096-4
https://doi.org/10.1016/B978-0-443-19096-4.00008-0

the influence of situational situations, oxytocin may have the opposite effect on them. Those who had an emergency delivery were much less likely to be angry and significantly more eager to encourage their children's exploratory behaviors than those who had a placebo delivery. During a talk concerning a disagreement between two people, it was revealed that oxytocin significantly increased positive communication behavior when compared to negative communication behavior (Pollack & Mayor, 2022). This finding suggests that the effects of this neuropeptide on social cognition have a neurological basis. When males were given oxytocin, their amygdala activity and connection to the midbrain decreased significantly. This offered proof that the effect was occurring. In terms of memory, intranasal oxytocin administration improved face recognition but not object recognition. It also increases sentiments of envy and schadenfreude. This is because studies have shown that oxytocin causes people to become more nationalistic. This neuropeptide is also thought to play a role in dishonest behavior carried out for the benefit of the group (Hammer et al., 2023, pp. 389—417). Depending on which area of the brain releases it, this hormone may have opposite effects on a man's aggressive tendencies. It has been stated that because of the significant disparities in biology and social structure, it is difficult to compare mouse aggression with human aggression. This is true even though both mice and humans use identical neurochemical and neuroanatomical pathways (Mierzejewska, 2022). Furthermore, there is a scarcity of research that demonstrates a clear link between the behavioral and neurological domains.

Researchers discovered that testosterone has a positive effect on both the stress response and individuals' socially aggressive and affiliative behaviors. As a result, giving women testosterone sublingually resulted in a significant improvement in the fairness of their bargaining behavior, which reduced the number of disagreements during bargaining and boosted the efficiency of social interactions. This was true regardless of whether the subjects were given testosterone (Nair et al., 2021). These unethical practises appear to be fostered by widespread misconceptions regarding testosterone's effects, such as the notion that the hormone causes people to become more antisocial, egocentric, or even hostile toward others. Furthermore, previous studies have shown that testosterone does not make people more violent; rather, it increases the likelihood of encountering aggressive behavior. It has been reported that male testosterone levels promote reactive aggression, whereas female testosterone levels do not. Instead, they propose that the hormone testosterone is the driving force behind a broader range of motivated activities, many of which are colloquially referred to as "dominance behavior." Most primates tend to exhibit nonaggressive primate dominance behavior, sometimes known as the urge to achieve or retain a dominant social position (Kashyap et al., 2022). Because the findings of correlational research have the potential to lead to inaccurate conclusions, they are normally handled with extreme caution. Breaking and entering, stealing, and vandalism are examples of these crimes. On the other hand, it has been noted that the direction of causality in this research is still unclear, as it is possible that the increased aggression was the source of the increased testosterone levels. This assumption stems from the fact that the direction of causality in these investigations is still unknown. In the past 10 years, significant progress has been made in the study of empathy using the SN approach. One's capacity for empathy can be defined as their ability to share the emotional condition of another person without mistaking it for their own (Nair et al., 2020). There are three primary manifestations of this behavior: experience sharing, which entails adopting the perspective of the target and experiencing their feelings and thoughts; mentalizing, which entails actively considering (and perhaps understanding) the feelings and thoughts of the target as well as the factors that contribute to those feelings and thoughts; and prosocial concern, which entails expressing the desire to work toward the improvement of the target's experiences. The inherent ambiguity of existing models of empathy may lead to interpretive uncertainty if the models are relied upon too firmly (Moini & Piran, 2020). This was found after the researchers analyzed the models' responses to a series of questions. It is the purpose of several of the proposed emphasis changes, some of which are gaining popularity among the next generation of empathy researchers, to avoid these and other errors.

The concept of "integrated levels" provides a comprehensive framework for understanding the evolution of matter from its most basic to its most sophisticated stages. This theory's purpose is to shed insight into the evolution of matter from its basic building blocks to its most sophisticated configurations (Sue Carter & Cushing, 2017). Matter can be organized in a variety of ways throughout its existence, from the physical to the chemical to the biological to the social. Simply described, emergence is the appearance of new structural and dynamic patterns in evolving matter inside complex systems. An alternative definition of "emergence" could be as follows: Social neuroscience is a multidisciplinary field that studies the biological underpinnings of social structures, processes, and behaviors, as well as the interactions. People pay close attention to the minute cues that are present in their immediate environment, which leads to this. To help achieve this goal, SN provides a standardized technique that can complete jobs on multiple levels (Krol et al., 2021). This strategy-based method is based on three determinism theories: multiple determinism, nonadditive determinism, and reciprocal determinism. Although there are some similarities between these theoretical notions and the concept of integrative levels, the goal of this research is to clarify the differences between the two so that it is clear what role each one plays in the whole.

Finally, SN demonstrates the possibility of interaction between biological and social components, implying that holistic knowledge from genetic to social levels of organization is required to fully explain social systems (Manukonda, 2018). A basic tenet of SN is the idea that some events can occur through the interaction of brain and social factors in ways that cannot be predicted by research conducted just in the disciplines of neuroscience or social psychology. This is the fundamental principle behind the study of SN. Second, the SN asserts that emergent phenomena are generated through increasing complexity inside a single level rather than by mixing variables from two levels, as the theory of the integrated level indicates. The concept of integrative levels, on the other hand, posits that merging variables from two levels leads to emergent phenomena (Carroll & James, 2019). This is because emergence occurs whenever the bar for integration at the same level is raised. This is since elements functioning at various levels cannot mix and produce new occurrences. Social behaviors and structures might be viewed as emergent phenomena rather than the consequence of integrating or combining social and biological components. That is because those characteristics are not the result of mixing those pieces. Emergent phenomena cannot be explained simply by focusing on a particular set of social and biological elements. This is because the combination of social and biological aspects does not result in fresh occurrences (Albrecht, 2021/2021). Given that, it is easy to see why this is the case. According to the integrated levels approach, there are distinct disparities between upward impacts that go from the biological to the social and downward influences that move from the social to the biological. There may be a distinction between the two, implying these distinctions. The reason for this is that emergent behavior is found in bottom-up systems rather than top-down processes.

Several organizational levels and isoforms have been postulated. To characterize the human brain, multiple layers of organization have been postulated for the neural circuits involved in information processing and cognitive and emotional functions. The notion foresees the formation of new integrated degrees of complexity at various stages of matter's evolution via various types of motion (Meenakshi et al., 2020). The pieces of a system that exist on various levels must be brought together to see the emerging features of the system. Despite the lack of solid proof, this theory has been advanced. As people learn about and use concepts like emergence, nonlinearity, attractive systems, and self-organization, their mechanistic and reductionist perspectives are altering. The emergence theory states that the properties of complex systems cannot be inferred from the properties of their constituent elements. This is because complex systems are made up of numerous simpler ones (Liu et al., 2021). However, the company's internal structure is only partially established. Nonetheless, the ontological status of levels is unaffected. The descriptive and integrative tiers are both considered here. Any proposed ontology should, at the very least, account for the physical-chemical, biological, psychological, and social dimensions. The emergentist paradigm allows for opposing ideas and points of view (Jung & Ryu, 2023). "Each organisational level reveals its own set of regularities that require distinct inquiry, just as each new level of occurrence requires its own specific attention and thought." "In the history of the universe, both basic laws and random events had a big impact on how matter came to be." The emergent system may be entirely consistent with physical principles. Even while biological development and evolution demonstrate a trend toward increasingly complex and differentiated states, the direction of events is toward "disorder" at all levels of organization. Even though things appear to be headed in the direction of "disorder," this is the case. Emergent layers do not break any physical laws and may even be in perfect agreement with them due to the additive nature of their underlying structure (Goddard, 2021). It is possible that the properties of living matter are crucial to everything from basic survival and protection to the discovery of totally new forms of life. It is difficult to find an argument that contradicts the overall logic. Just as there are physical and biological laws that only apply at certain levels of organization, there are laws that govern the overall evolution of matter over its entire lifetime. These latter laws ensure that society functions properly. We call a process "rising" or "emergence" when a more complex phenomenon emerges from a simpler one. When effects propagate from lower to higher levels of an entity, we say that "upward causation" has occurred. Therefore, we may be able to attribute the effect observed at the higher level to the lower level. Downward causation exemplifies how emergent events can modify the properties of subsystems. A feedback loop between emergent phenomena and deeper layers could have been the source of this.

It is possible that separate substances are not always required for there to be distinct organizational levels. This is because the material composition remains constant as more levels of complexity are added. This is because chemicals can take on a variety of structural forms. There is only one physical planet, but it is not without variety. Some interpretations of DC indicate that higher-level entities do not advance by associating with lower-level entities; however, others argue that this is contrary to the scientific principles that underpin DC. Both schools of thought begin with the premise that connecting two separate entities working at different levels produces no positive consequences. According to the concept of strong direct causality, higher-level entities exist independently of lower-level ones. This philosophy is based on the principle of constitutive irreductionism. Overall, the integrative levels present a picture of a material cosmos that is not only tremendously complex but also hierarchically structured. To solve the mind versus body conundrum, it is critical to remember that the mind is not a separate thing but emerges ontologically from the interaction of the brain, the body, and the surrounding environment. This is a truth that cannot be overlooked. According to the most extreme form of the reductionist perspective, all mental states, including consciousness, are thought to be explainable in terms of biochemical processes occurring in the brain. The threat increases when the data analyze the actions of living organisms. This is because when more specific information is provided, greater conclusions can be drawn from the data. This occurs because the complexity of the appropriate explanations increases as more information is discovered. It is likely that scientists are not looking in the correct places for the chemicals that allow us to learn and interact socially. In their effort to understand the inner workings of the brain, researchers are most likely seeking these compounds in the incorrect area. Because humans have a proclivity to seek explanations for functional activity at lower integration levels, we may miss the appearance of emergent features (Kelly & Goodson, 2014). When we experience challenges, we tend to hunt for causal elements in lower-level processes. When scientists attempt to "scale up" or extrapolate evidence from overly simplistic models of complex psychological phenomena, major issues may develop. Given these possibilities, it may be more challenging to find solutions to difficulties. Extrapolating from the results of these models is notoriously difficult for academics. They argue that the events under consideration are so complicated that any attempt to present a reductionist explanation will fail. Because each of these scenarios poses unique challenges, even though the mind is obviously dependent on the brain, it is impossible to divorce the mind from the brain. As a result, it is theoretically impossible to separate consciousness from brain function. Even though the mind is dependent on the brain, some people argue that it is not physically located in the brain. Although nondualists do not believe in a unique immaterial essence behind these mental processes (hence the term "nondualism"), they do seek to reject orthodox mechanical materialism, which claims to reduce higher-order phenomena to lower-order phenomena. Nondualists hold that higher-order phenomena can never be reduced to lower-level phenomena. Mechanical materialists promoted a sort of reductionism that was popular in the nineteenth century. Some people believed that awareness was merely an unintended result of how the brain worked. The fundamental goal of this viewpoint is to present an alternative to the prevailing mechanistic materialist worldview, which incorrectly argues that all events have a single, overarching cause.

2. Related work

The human brain is said to be a complicated system because mental states are created through the collaboration of many different physical and functional levels. The human brain is far more complex than any other structure that is currently known. According to this school of thought, the brain, body, and environment are all mutually embedded systems that are intricately entwined on multiple levels of connection. This occurs because the human brain is an intrinsic component of the layers that constitute the human's surroundings. As new brain processes emerge, they can influence change by feeding back to more fundamental systems, a phenomenon known as dynamic coupling. As a result, the transition from interpersonal to intrapersonal processes would necessitate a convoluted series of experiences and the internalization of activities with deep social roots and a long evolutionary history. Such instruments include languages, counting and writing systems, algebraic symbol systems, and other similar systems. This category, however, is not limited to just these. Vygotsky's explanation for the acquisition of advanced cognitive capacities relies heavily on the variety of meditative activities that occur during development. When Vygotsky alluded to "social" processes in his books, he meant "face-to-face," dyadic activities. He mislabeled them, but what he wanted to say was "inner psychological processes." Despite this, he acknowledged the existence of a deeper level of social phenomena, which Wertsch referred to as "social institutional" processes (Kashyap, 2019). To imply that inner psychological processes can be reduced to those occurring within a single mind is a sort of psychological reductionism. In a similar spirit, Wertsch claims that when one reduces societal processes to those of interpersonal psychology, a sort of reductionism is present.

Educational institutions have no choice but to transition away from their dependency on physical campuses and toward online classes. Most countries have never even heard of anything remotely like primary and secondary education that could be completed entirely online. To ensure that the current academic year would proceed without a hitch, the educational community invested. Because of this, the entirety of the time spent providing the urgent online instruction was characterized by the utilization of extremely rapid processes. Students and teachers reacted to the current situation and produced responses that are conducive to the successful implementation of the construction operation. This was accomplished despite the hurdles that were presented to them. This method of processing was also required for classes that were attended at technical universities by students who were already in their mature years. Students currently enrolled in the computer science curriculum at vocational high schools will be the only ones permitted to participate in the research. Because of this, the research is severely limited in its capacity for exploratory analysis. This education will cover not only the theoretical but also the practical aspects of the field. The findings of this study will be compiled and analyzed with the goal of gaining a better understanding of the opinions held by adult students regarding the topic of online education (Doukakis & Alexopoulos, 2020). Technical upper high schools are an integral part of the Greek secondary vocational education system, which would not be complete without them. These educational establishments provide instruction in several specialized fields. Classes at vocational high schools might take place either during the day or at night, depending on the schools' own timetables. Evening programs are convenient for students who are currently employed but would still like to complete their secondary education. The following is not an exhaustive list of the vital roles that vocational upper high schools play in their respective communities, but they are among the most important. Educating young people about the history and culture of Europe, with a particular emphasis on the preservation of ethnic, religious, and cultural traditions, and preparing them to become contributing members of society in Europe students at the vocational high school can participate in both the "upper secondary education cycle" as well as the "vocational high school diploma course." Within the scope of this article, the "postsecondary cycle" and the "apprenticeship cycle" are interchangeable terms for the same thing.

One of these fields is known as informatics, which is also known as computer science. Informatics application technicians and computer and network technicians are just two examples of the kinds of work that might be found in this field. There are many other types of jobs that could be found here as well (Tsiara et al., 2019). Students in both traditional high schools and vocational high schools who successfully complete all the necessary criteria for graduation are awarded either a diploma that is comparable to that of a traditional high school or a degree of specialization at level 3. Apprentices who successfully complete all the prerequisites necessary to have their credentials validated are rewarded with level 4° as a sign of appreciation for their hard work and dedication. In the last phase, students who have graduated from high school are required to take standardized examinations that are given on a nationwide scale to determine whether they are academically prepared for the subsequent level.

3. Training and academic prerequisites

Disaster education and planning utilizing a satellite for many years, academic institutions such as universities have used and perfected the "e-learning" teaching approach. Institutions with the means and know-how to implement remote learning are the best candidates for conducting early pilot projects. To meet the unforeseen needs brought on by the epidemic, communities and school administrations have turned to online training. This shift was brought about by the increased emphasis placed on achieving educational goals. On the other hand, timely education was critical for the academic community. Even though much time and effort were put into planning and preparation, properly implementing this strategy proved tough. This form of instruction is now referred to as "urgent situation remote teaching" to differentiate it from "distance education," which, as previously said, comprises both broad concepts and more specialized methods. In theory, if this option had been chosen, more individuals would be able to afford college.

Web-based teaching approaches can help meet some of the goals of remote education delivery. These goals may include providing instruction remotely or during times of calamity. Synchronous distance learning allows teachers and students from different locations to collaborate in real-time on group assignments and chats. The term "live streaming" is used to characterize it in the context of remote learning (Ansari et al., 2021). This update welcomes both sides' communication and comments. As a result, the coach and the student can be in different locations at the same time while still monitoring the session's progress. Students enrolling in an asynchronous online course do not need to be in the same location or even the same time zone to engage in discussions and other activities. The student must decide when to start working hard in school. Teachers used remote education in both its asynchronous and synchronous versions to reach as many pupils as possible during the early stages of the epidemic. We did this to keep the sickness from spreading further.

Long-distance learning has received a lot of attention. There has been an increase in research and publications in response to the epidemic and the rise of online schooling. This is because the number of papers and articles published in recent years has increased significantly. Recent investigations have shown a slew of issues with online learning. This is necessary in order for the school to provide a high-quality education to all of its students. According to current research, the adoption of COVID-19 has significantly altered online learning. According to the data analysis, the most prevalent problems for the trainees were (a) technology, (b) the trainees' technical and knowledge abilities, (c) the educators' efforts to enhance the learners' motivation, and (d) the trainees' concerns about a lack of resources. At this stage in the conversation, citations are required. The conclusions of a second poll, which sought to ascertain how instructors and students alike felt about the use of mandatory online attendance in the aftermath of the pandemic, were consistent with the findings of the first study. We sought to gain insight into both teachers' and students' perspectives by completing this survey. Respondents to the study stated that they saw no evidence of lecturers or their students straying too much from the set lecture format. Several interviewees mentioned this during our discussions. Since I left, my overall sense of contentment has decreased. Despite this, it was obvious that after frequent, thorough, and timely contact was introduced into the teaching process, both the instructors and the students were satisfied with the online courses. It is critical for instructors and students to have a good attitude during the shift from obligatory remote learning to distance education. Knowledge integration solutions have demonstrated efficacy comparable to distance education in the realm of vocational training area with a variety of laboratories for students to use for customized coursework. According to a survey of 508 people in vocational training, participation in distance education had a significant impact on students' ability to interact. However, according to one study, most people are unaware that distant learning improves cognitive performance. In any event, male respondents were more positive about distant learning than female respondents. According to a more recent poll of 206 vocational school students, most pupils appear to be opposed to using online teaching approaches in language classrooms (Kashyap, 2019). Recent research, however, shows that most students have no problems using online language learning programs. Despite this, data show that if given the option, most students (74%) would choose to finish their assignments and take tests online. Because of the importance of memory in the development of educational neuroscience, both the final test and the intermediate evaluation are aimed at improving students' recall of material. The structure, content, and activities of the course all influence how students evaluate the success of distance learning. This is because increasing each of these three factors has a multiplier effect on the overall effectiveness of the educational process.

Similar research on students participating in tourism-related vocational courses revealed that the vast majority had access to smartphones with relevant educational apps, allowing them to spend extended periods of time studying online. This conclusion was made based on replies to a poll about specific majors. The increasing use of portable technology enabled the discovery. The kids declared that they would not need any more help from others. They insisted on finishing their studies online, even though they were cut off from their instructors and classmates for the duration of the pandemic. Despite this admission, research shows that children do best in classes where they can cooperate to learn. It is commonly understood that employing a hybrid approach to teaching vocational topics offers benefits. Students believe that the most effective educational technique is to combine regular classroom instruction with online learning technologies such as Moodle. That will be the beginning of another problem. To summarize, the topic of distant labs is one of the most critical to address when considering vocational training. According to recent research, allowing students to use the facilities and keep track of their own data allow them to undertake further analysis and increases their grasp of technical concepts. The researchers arrived at this result after ensuring that all students had adequate access to relevant information and were proficient in data management. The findings revealed that with the correct approach to learning, instructional materials, and supporting material, students may have fun while learning in the lab. Based on the findings thus far, it would be good to investigate how educational neuroscience may be applied to online learning and how its principles might be included in carefully planned activities to improve the quality of the educational experience for distant learners. As will be mentioned in the following paragraphs, cognitive psychology can be applied in the classroom and to create works of art.

4. Educational neuroscience

Educational neuroscience stands out as a shining example of a successful model for interdisciplinary research because it brings together researchers from several academic fields to collaborate with teachers on the development of cutting-edge instructional practices. To make decisions and move forward, the manual based all its decisions and advancements on reliable, objective data. We hope that by doing so, we will be able to help students: (a) become more invested in their education; (b) generate creative ideas; and (c) "store" their content in a way that will allow them to access it in the future. The implementation of the evidence provided in this article is expected to result in the following outcomes: This research has resulted in the creation of educational neuroscience as a new field with the potential to transform both formal classroom instruction and the educational landscape. This is made possible because educational neuroscience brings together scholars from a wide range of disciplines. The goal of this field is to profoundly transform the way people learn. These early studies played a crucial role in laying the groundwork for what is now known as educational neuroscience.

This article discusses numerous ways that educators can employ to promote critical thinking and knowledge growth in their students. Several strategies are explored in the essay. Furthermore, they allow children to learn in an environment that supports the development of skills such as the ability to think critically, perceive, recall material for extended periods of time, and pay close attention. Teachers can use a variety of strategies to keep students interested in the material they are studying. Code words, small-group dialogue, putting information on electronic whiteboards, and allowing temporary group separation are all examples of comparable tactics. For your convenience, we have highlighted some of the numerous strategies that the instructor may employ. When the instructor aids, the method's principal goal encouraging student participation is considerably more likely to be met.

Digital diagnostic, formative, and summative assessments benefit students by allowing them to consolidate, retain, and recall newly learned knowledge more quickly. The instructor has two options: critically analyze their own techniques of training or assess their students' development in whatever way they feel is appropriate. Students can improve their memory, demonstrate what they have learned, and express their needs by voting on right or wrong type questions as well as creating questions with multiple choices or equal consequences during the lecture or as homework. Voting on right-or-wrong questions can also help students demonstrate what they have learned (in an asynchronous context). This purpose can be met in a variety of ways, including by employing yes/no questions for in-class polling, writing questions during instruction, and using questions for assignments (in an asynchronous context). Students can influence the design of multiple-choice questions as well as how incorrect answers are treated. However, when assessment is based on exciting ideas in addition to the individual talents and qualities of specific students, children are more engaged, and opportunities for differentiation are formed [26]. Working with teachers to address any learning issues that have been identified in their children is one way that can be used to enhance a child's brain's healthy development. Students must actively participate in the learning process to profit from receiving an education online. One way to accomplish this goal is to respond to questions or statements made by the teacher or other students in the classroom. To attain this goal, students will have the opportunity to conduct mistake analysis both individually and in smaller groups. Individuals that do not comply with accepted norms in their cognitive processes or activity styles fall into the fourth category. Educators that use this method will be able to provide students with access to a wide range of conceptual frameworks. Furthermore, it aids in the general growth of their brains, strengthening the connections between the various sections of their brains. According to the field of educational neurobiology, the two most significant conditions for adopting any of the four tactics are a focus on the learner's neural architecture and an understanding of how the learner's brain processes. Four online courses must be able to continuously commit time to studying to excel academically. It is critical that children spend this time alone since the process of forming new synaptic connections and circuits in the brain takes so long. You should not spend it attempting to develop solutions for automating operations or duplicating data. Working with others to reach a common goal is vital, but it is also a character-building exercise in humility and is critical to one's evolution as a learner, employee, and thinker. Students benefit from collaborative problem-solving in two ways: (a) it helps people get better at addressing the issue at hand, articulating their concerns and suggestions to one another, and acting and (b) it helps them become more aware of and grateful for each team member's individual contributions. These two benefits are critical for students' development. Giving students the option of using the breakout rooms is one of several techniques for encouraging them to engage and collaborate with one another. Teachers can tailor their classes to their students' requirements to collaborate with them on group projects. Right at the start of the survey administration procedure, respondents were required to supply some basic personal information. In the second part of the survey, which was made up of 14 open-ended questions, students had the chance to say what they thought were the pros of getting an education online.

In the words that follow, each of the survey's first 14 questions will be thoroughly analyzed.

Before enrolling in any online courses or programs, make sure you have the essential hardware and software. This category includes all electronic gadgets that can connect to the Internet. Have you gotten all the materials required to finish the project? Have you ever encountered a technical issue while attempting to complete an online course? Have you, for example, had trouble accessing the lesson because your internet service was down or your electronic device ran out of power? If this is the case, it would be extremely helpful if you could describe the steps you took to resolve the issue. How confident are you that you can access a web-based platform such as WebEx to participate in a scheduled activity? Please keep me updated on the status of this request, if possible. Do you think online learning can meet your needs right now, or would you rather go for asynchronous training? Do you believe you learn more efficiently in a classroom setting when you follow a teacher's directions or when you use a free Internet resource? Students who choose to pursue their education in a virtual setting will be exposed to an increasing number of cutting-edge concepts and ideas throughout their studies. Is there another course of action you can pursue now that you have been treated with this therapy? Have the students demonstrated that they understand the topic and have the relevant underlying knowledge by participating in the online course? Most educational institutions require students to understand the proper safety precautions to take in a laboratory setting as part of their academic program. How far along are we in using online education to achieve this goal? When you are not in the same room as your instructors, how do you communicate with them? Did their friends devote the same amount of time and effort to both academic and physical education? Do you have a plan for how you spend your time, such as how you study, clean, and exercise? What are your thoughts on taking courses online rather than in person? What are the benefits and drawbacks of taking classes online, and how does it compare to the more traditional classroom setting? Is it more difficult for you to build meaningful relationships with other kids now that you must go to school every day? The children were highly encouraged to answer all the questions honestly because doing so was essential to creating an accurate record of the students' viewpoints. The questions and answers were all kept strictly confidential, and no information was shared with anyone.

5. Findings

The investigation was carried out between 2020 and 2021. Participants in the study were given the option of remaining anonymous, and the data gathered were kept private to ensure its confidentiality. Case studies are drawn from a wide range of real-life scenarios and topics. Every single high school participant was enrolled in at least one of their own institution's after-school vocational programs. The investigation involved 10 people in total, five of whom were men and five of whom were women. When it comes to high school graduation, seven graduates from the advanced track are equivalent to one graduate from the regular track. Only three of the nine students intend to continue their education after high school; the remaining seven are completely committed to becoming industry leaders in information technology. Everyone who enrolls in the program does so because they are dedicated to making the most of their time there and are serious about obtaining a master's degree in their chosen field. In the morning, five students are given the opportunity to gain experience in a professional setting. Summer jobs are available for three college students. There are two current students at our school who are looking for work. The information gathered about the population is divided into the following groups or subsets: A 29-year-old single man was able to get a postsecondary credential even though he worked as a seasonal worker during the busiest time for the tourism industry. A man who is at least 31 years old, has completed an advanced high school program, is married, has at least one child, and works in the private sector. In addition, he must be the father of at least one child. A woman with a high school diploma who has been married for 34 years is a third example. This individual has not demonstrated academic success. A married woman in her forties is the fourth example. She now has a significant degree and is employed in a significant field because of the work she did in high school. A 47-year-old married man with an associate degree works in the private sector. Number six on the list is an unmarried 43-year-old woman. She has a high school diploma and works in the private sector. She is a married woman in her late seventies with no children. She has an associate degree and works seasonal jobs in a variety of industries to gain experience. A man in his forties who works in the private sector is married but has no children. A married man over the age of 27 has a high school diploma and has taken advanced placement courses. A 36-year-old male with a bachelor's degree in education or its equivalent provided the participants' responses. The analysis and interpretation were completed with the knowledge gathered from the study's data. Personal information from participants was collected at the same time as the census. This goal was the focus of all efforts made during the data collection process. The discussion that follows will cover both the quantitative and qualitative aspects of the study's findings. The previously mentioned events will occur, followed by a discussion of the study's findings. The start of a suspicion six of the students reported having all the necessary components on hand and having no problems with the technology. As a result, using digital technology is entirely appropriate. Specifically concerning the second question, seven students reported minor issues such as a power outage or difficulty logging on to the Internet. Only three students claimed to have had no problems attending their classes. According to the responses, it appears that resolving the technical issues will be a difficult task. However, because of these issues, training is frequently hampered in an ineffective manner. The third and most obvious follow-up question is: "Despite their advanced degrees, the respondents demonstrated that they were more than capable of overcoming the unique academic challenges that distance learning presented." The picture that emerges from the responses is uncannily like the one presented in the initial inquiry. To put it another way, it appears that students who are actively engaged in their studies have attained a level of technical proficiency that allows them to effectively use mobile technology. This is because these students are focused on their studies. Please respond to the following question: One of the adults who took part in the study said they would be open to a format in which theory was delivered digitally, and labs were held in predetermined locations. Three of the adults who took part in the study mentioned the growing demand for online education. Six of the adults who took part in the study said they would take the class if it was held in a more traditional classroom setting. Adults appear unconcerned about beginning their college studies online, and it is possible that they are even eager to get started. We were able to conduct in-depth research on this topic and zero in on the most important aspects of the application process with the help of the qualitative research tool NVivo. The following section will lay the groundwork for our in-depth investigation. Fifth, let us consider one more point: Many online students have stated that they would prefer to participate in a class that meets at a specific time rather than physically attending a class that meets at various times. When we consider that we are talking about adult learners, the answers to this question take on a whole new level of mystery and intrigue. The sixth question can be addressed after this, so let us get started. Educators and policymakers at the Department of Education issued a circular instructing teacher to review previously covered material in online courses rather than introduce new ideas during the 2019—20 academic year. This was supposed to be finished before moving on to the next piece of content. This updated curriculum will be introduced during the 2019—20 academic year. This task needed to be completed as soon as possible to meet the requirements stated in the instructors' directive. The instructors assigned to the online education program were tasked with continuing in the same manner as in previous years and ensuring that all pertinent information was covered during the upcoming school year of 2020—21. Given the new information, it is reasonable to wonder whether today's students are prepared to deal with this novel development. When enrolling in online courses, students frequently express concern about their ability to keep up with the material and meet the course's learning objectives. Six students raised their hands to show that they had completed the lesson objectives and had a thorough understanding of the subject. Four of the students, on the other hand, have complained that they are not making enough progress in class. The question has apparently been posed to the group eight times now. It will be fascinating to hear the participants' perspectives on how they spent their time in the lab, so bring your questions. A computer science laboratory may specialize in either hardware or software. Two students were pleased with the progress made in laboratory lessons; six students reported having difficulty getting to the labs, and two students are keeping an open mind because they are unsure how well they achieved the desired learning outcomes. In the ninth investigation, the solution to your problem will be revealed. The answers to the questions posed by the respondents are provided in the paragraph that follows this one. Three more students stated that they had no reservations about their teachers having online access to them, but they preferred speaking with their teachers in person. Even though they did not say they were worried, these students said they liked talking to people in person. Finally, on the question that I believe is most important, 80% of those who responded to the survey agreed with the statement that students may be more engaged in a lesson delivered online rather than in a traditional classroom. The study's findings revealed that the participants held these views. One student took a position on the issue that was diametrically opposed to the other students, and the other student took no position at all. In the 11th round of questions and answers, one of the interesting challenges that adults who return to school face are trying to fulfill the educational commitments they made to themselves prior to beginning their studies. Eight students said they had been very consistent in their roles as educators in the context of e-learning, while only two said they had not. One side claimed that the other was consistent because he lacked the necessary technological resources, but both sides accused the other of attempting to avoid their respective obligations. Both parties accused the other of attempting to avoid their respective obligations (he could only use a tablet for most of his work). One camp claimed that he was unable to complete the task because he lacked the necessary supplies (he could only use a tablet for most of his work). The problem with asking 13 and a half questions is that there are not enough answers. Students' thinking patterns appear to have been significantly influenced by their online education. Eight students claimed that the courses' solely online nature had no impact on their worldview, while two claimed the opposite. This was in response to a question about the impact of attending an entirely online school on students' psychological health. The 13 distinct concerns students found it difficult to stay in touch with one another and form new friendships due to communication restrictions imposed by the pandemic. While the other four students claimed they did not miss talking to their classmates because they already did, six students claimed they did not miss it because they already did. In the 14th question, participants were asked to weigh the benefits and drawbacks of getting their education online in a free-response question. In the "positives" column, we listed the students' homes, their perceptions of personal safety, and the educational opportunities that were made available to them. Some problems have been brought up, such as the low quality of the online world and the problems with correlation. As a result, students were able to learn, study, and advance in both their academic and personal endeavors. Researchers found that students were more focused in classrooms that were set up to help them succeed. This made students less anxious and made them more productive overall.

6. Importance of social and educational neuroscience

We ran additional analyses on the data using the NVivo software and discovered some intriguing new findings. Learning might change if they were required to take online courses from the comfort of their own homes. In this scenario, students could participate in class discussions and have access to all course materials. Students can review their online study sessions whenever they want, from the comfort of their own homes. Many students prefer to complete their schoolwork in solitude at home for the reasons depicted in Fig. 4.1. Some students, particularly those with families, may find that studying at home is more convenient. Staying put also provides increased security and a lower risk of contracting infectious diseases such as COVID-19. These benefits are more important than not having to move and not having to deal with the hassle and cost of packing and moving.

Women were more optimistic than men about the opportunities that could be realized through distance education. Married women, for example, are increasingly interested in online education because it allows them to support their families while also furthering their education. This is one of the factors influencing the increased popularity of online education. M4, the group's youngest and most active student, is also the most optimistic about the prospect of completing their degree entirely online. M4 works a full-time job while also attending school. The Education Board's online education platform was clearly the root of the problems, as most criticisms directed at it can be traced back to it. Based on these results, it seems that the platform is the main reason why online education does not work.

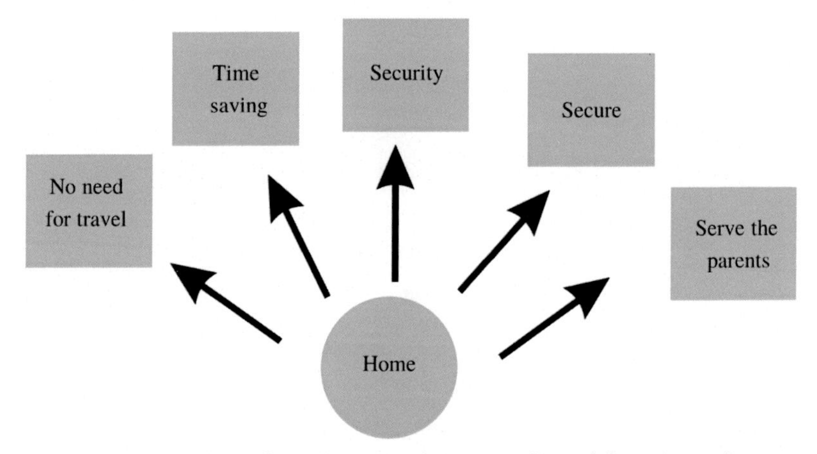

Figure 4.1 Arguments for getting an education online while sitting at home.

When asked about the potential benefits of online education, female students responded more positively than male students. For example, married women are increasingly turning to online classes to further their education while still providing for their families. As a result, married women are taking more online courses. This is one of the many reasons why married women are interested in online education. This encourages more people to enroll in distance learning programs, which in turn encourages more people to enroll. M4 is the newest, liveliest, and most optimistic member of the crew. They are also the most optimistic about the possibility of earning their degree entirely online. M4 works full time at a job of his choice in addition to his academic pursuits. Given that most of the complaints filed against the education board can be traced back to its online education platform, this is where most of them began. This clearly shows that most of the complaints filed against the education board were about its online education platform. Based on these results, it seems likely that the platform used to deliver online education is the main reason why it does not work well (see Fig. 4.2).

The opinions of the pupils as they contrast learning online with studying in the actual school setting and explains the reasons why they want to have been in the actual classroom setting. The ability to get their education online is beneficial to them for both career and familial purposes. Finally, the numerous factors that influence students' decisions to choose synchronous distance learning over asynchronous distance learning when given the

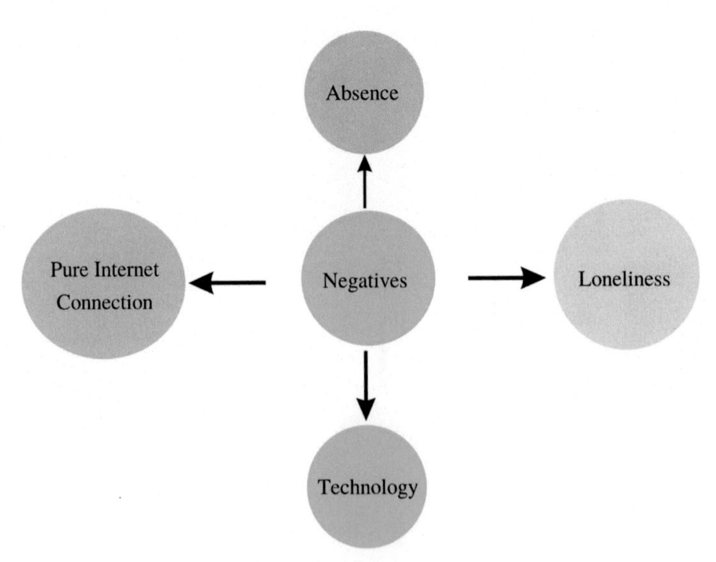

Figure 4.2 Reasons against online education.

choice. If the tree is any guide, the use of concurrent e-learning allows for the restoration of communication, which had been largely eliminated from everyday life due to the epidemic. This was a significant change from before the epidemic. The "mandatory" nature of student involvement in the asynchronous course appears to be derived from the course's structure. The seventh section includes some summaries. According to the findings, students are open to the idea of furthering their education through distance learning for two main reasons: (a) it allows them to do so from the comfort of their own homes and (b) it protects them from exposure to harmful software. Both factors influence students' willingness to pursue their education via distance learning. Married women and those who have completed high school are the demographics most likely to favor online learning. This is an added benefit. They highlight one of the teaching platform's drawbacks, which is that maintaining a connection can be difficult at times. However, students believe that natural learning environments cannot be replicated because of the close relationships and open communication that exist between teachers and students. Despite this, many people believe that synchronous online learning can benefit them both personally and professionally. Aside from the infrequent issues caused by technical difficulties, students have generally had no problems with the course material. As a result, it appears that the information is presented conventionally and that the students can effectively absorb what they are taught. Based on their responses, adult students appear to believe that taking classes online is safer, despite the possibility of complications caused by the pandemic. It looks like one of the most important ways to get past problems is for students and teachers to work well together. According to the students' overall description of the classroom experience, the teachers created a very supportive environment, and the collaborative efforts of the teachers and students were mostly related to educational neuroscience principles. This is supported by the students' accounts of their classroom experiences. To put it another way, it appears that conversations between students as well as between teachers and students had a positive impact on the development of brain connections. Furthermore, it appears that the evaluation was designed primarily to serve a formative function and that the environment in which the student was evaluated was not intimidating. People were thus strongly encouraged to participate in the activity. The data do not support the use of learner-provided instructional barriers or a multidimensional approach to instruction. Using instructional barriers provided by learners is also inappropriate. When students are asked to participate in laboratory classes, they are put to the test because the material covered there is difficult and requires additional preparation. A framework for collaborative learning has been developed, which will be extremely beneficial to the student's intellectual development. During the pandemic, the study provided some novel insights into the adult education resources that were available online. Making the most of the numerous positive effects produced by online education is critical when education returns to more traditional classroom settings.

7. Conclusion

Understanding integrative organisational levels are thus required to comprehend how the inanimate, living, and social worlds interact. Despite this, predicting what will happen on a higher level based on past occurrences can be difficult. The links between the two levels are simply too strong. This suggests that to learn the principles that are unique to that level, one must adopt tactics appropriate for that level. To put it another way, one must compete at the appropriate level to thoroughly understand the rules. To begin, SN believes that the union of brain and social components might result in emergent phenomena, which are observable events that cannot be predicted by normal neuroscientific or social psychology techniques alone. Emergent phenomena are observable events that cannot be predicted using traditional social or neuroscientific methods. However, according to the assumptions of standard integrative analysis, emergent phenomena are the result of an increase in complexity at a single level rather than the integration of variables from two different levels. Despite the probability that this is the case, it appears that there should be more than one comment made about this issue. When conducting animal research, it is critical to avoid the mistake of focusing primarily on simple behavioral assessments rather than more complicated behaviors. This is one of the most common mistakes in this line of business. Furthermore, most of the research in the field of human neuroendocrinology focuses on inner psychological phenomena. These phenomena include "face-to-face," "dyadic," and "small-group" communicating processes.

References

Albrecht, J. (2021). Coda: Charting future directions of music cognition in turbulent times. *Future Directions of Music Cognition*, 270–276. https://doi.org/10.18061/fdmc.2021.0054 (Original work published 2021).

Ansari, G., Garg, M., & Saxena, C. (2021). Data augmentation for mental health classification on social media. *arXiv*. https://arxiv.org.

Cacioppo, J. T., & Decety, J. (2011). Social neuroscience: Challenges and opportunities in the study of complex behavior. *Annals of the New York Academy of Sciences, 1224*(1), 162—173. https://doi.org/10.1111/j.1749-6632.2010.05858.x

Cacioppo, J. T., Hawkley, L. C., Norman, G. J., & Berntson, G. G. (2011). Social isolation. *Annals of the New York Academy of Sciences, 1231*(1), 17—22. https://doi.org/10.1111/j.1749-6632.2011.06028.x

Carré, J. M., & Olmstead, N. A. (2015). Social neuroendocrinology of human aggression: Examining the role of competition-induced testosterone dynamics. *Neuroscience, 286*, 171—186. https://doi.org/10.1016/j.neuroscience.2014.11.029

Carroll, H. A., & James, L. J. (2019). Hydration, arginine vasopressin, and glucoregulatory health in humans: A critical perspective. *Nutrients, 11*(6). https://doi.org/10.3390/nu11061201

Doukakis, S., & Alexopoulos, E. C. (2020). Knowledge transformation and distance learning for secondary education students. The role of educational neuroscience. In *SEEDA-CECNSM 2020 - 5th South-East Europe design automation, computer engineering, computer networks and social Media conference*. Institute of Electrical and Electronics Engineers Inc. https://doi.org/10.1109/SEEDA-CECNSM49515.2020.9221821

Goddard, K. E. (2021). Consequences of an obesogenic diet can be prevented by knockout of P2Y6 purinergic receptor in mice. *Purinergic Signalling, 17*(3), 323—325. https://doi.org/10.1007/s11302-021-09793-8

Hammer, Leslie B., Allen, Shalene J., & Leslie, Jordyn J. (2023). *Occupational stress and well-being*. Cambridge University Press (CUP). https://doi.org/10.1017/9781009268332.017

Jung, Y. A., & Ryu, J. (2023). Associations between obesity and academic enthusiasm and social emotional competence: Moderating effects of gender and sleep quality. *Journal of Human Behavior in the Social Environment, 33*(2), 276—295. https://doi.org/10.1080/10911359.2022.2052224

Kashyap, R. (2019a). Big data Analytics challenges and solutions. In *Big data Analytics for intelligent Healthcare management* (pp. 19—41). Elsevier. https://doi.org/10.1016/B978-0-12-818146-1.00002-7

Kashyap, Ramgopal (2019b). Security, reliability, and performance assessment for Healthcare Biometrics. *IGI Global*, 29—54. https://doi.org/10.4018/978-1-5225-7525-2.ch002

Kashyap, R., Nair, R., Gangadharan, S. M. P., Botto-Tobar, M., Farooq, S., & Rizwan, A. (2022). Glaucoma detection and classification using improved U-net deep learning model. *Healthcare, 10*(12). https://doi.org/10.3390/healthcare10122497

Kelly, A. M., & Goodson, J. L. (2014). Social functions of individual vasopressin-oxytocin cell groups in vertebrates: What do we really know? *Frontiers in Neuroendocrinology, 35*(4), 512—529. https://doi.org/10.1016/j.yfrne.2014.04.005

Krol, K. M., Namaky, N., Monakhov, M. V., Lai, P. S., Ebstein, R., & Grossmann, T. (2021). Genetic variation in the oxytocin system and its link to social motivation in human infants. *Psychoneuroendocrinology, 131*. https://doi.org/10.1016/j.psyneuen.2021.105290

Liu, C., Song, Z., & Shi, R. (2021). Neural bases of brand reputation effect on extension evaluation: An ERPs study. *Frontiers in Neuroscience, 15*. https://doi.org/10.3389/fnins.2021.704459

Manukonda, Veera (2018). New developments in autism treatment. *Autism-Open Access, 08*(03). https://doi.org/10.4172/2165-7890.1000e143

Meenakshi, P., Kumar, S., & Balaji, J. (2020). Immediate early gene expression dynamics in vivo segregates neuronal ensemble of multiple memories. *bioRxiv*. https://doi.org/10.1101/2020.12.17.423270

Mierzejewska, W. (2022). Cooperation, competition, and coopetition within business groups. In *Business groups and strategic coopetition* (pp. 22—40). Taylor and Francis. https://doi.org/10.4324/9781003324775-3

Moini, J., & Piran, P. (2020). *Embryology* (pp. 51–76). Elsevier BV. https://doi.org/10.1016/b978-0-12-817424-1.00002-1

Nair, R., Singh, D. K., Ashu, Yadav, S., & Bakshi, S. (2020). Hand Gesture Recognition system for physically challenged people using IoT. In *2020 6th international conference on advanced computing and communication systems, ICACCS 2020* (pp. 671–675). Institute of Electrical and Electronics Engineers Inc. https://doi.org/10.1109/ICACCS48705.2020.9074226

Nair, R., Vishwakarma, S., Soni, M., Patel, T., & Joshi, S. (2021). Detection of COVID-19 cases through X-ray images using hybrid deep neural network. *World Journal of Engineering, 19*(1), 33–39. https://doi.org/10.1108/wje-10-2020-0529

Pollack, S., & Mayor, C. (2022). The how of social justice education in social work: Decentering colonial whiteness and building relational reflexivity through circle pedagogy and Image Theatre. *Social Work Education*, 1–16. https://doi.org/10.1080/02615479.2022.2104244

Sue Carter, C., & Cushing, B. S. (2017). Proximate Mechanisms regulating sociality and social monogamy. In *The context of evolution* (pp. 99–121). Informa UK Limited. https://doi.org/10.4324/9781315133676-8

Tsiara, A., Mikropoulos, T. A., & Chalki, P. (2019). EEG systems for educational neuroscience. In *Lecture Notes in computer science (including subseries lecture Notes in artificial intelligence and lecture Notes in Bioinformatics)* (Vol 11573, pp. 575–586). Springer Verlag. https://doi.org/10.1007/978-3-030-23563-5_45

CHAPTER FIVE

Emotional AI: Computationally intelligent devices for education

M. Keerthika, K. Abilash, M. Sundararaj Vasanth, M.K. Kuralamudhu and Subramaniam Abbirooban
Department of Psychology (SF), PSG College of Arts and Science, Coimbatore, Tamil Nadu, India

1. Introduction

The integration of emotional intelligence into machines like robots, chatbots, and virtual assistants is the focus of the emerging and fascinating field of artificial emotional intelligence (AEI). The development of AEI has made considerable strides in recent years, and it is now a vital part of human-machine interaction. The idea of AEI, its significance, and how it is being applied will all be covered in this chapter. The needs in the educational setup are growing too fast and the emergence to fulfill those needs is a necessity. A country's youngsters need to receive good education in order to be better prepared for life. Instruction in academic topics alone does not constitute proper education, nor can it be separated from the child's emotional needs. Academic progress is largely affected by emotional intelligence (EI). There are enormous studies that have shifted the focus toward emotional intelligence of educators. Teacher's job involvement may act as an intermediary between the connection between teacher emotional intelligence and student academic progress (Wang, 2022).

The ability to understand, use, control, and regulate emotions is the most typical definition of EI. People with high emotional intelligence are able to recognize their own feelings as well as those of others, apply emotions to guide their decisions and actions, distinguish between various emotions and give them the appropriate names, and adjust their emotions in reaction to their environment (Colman, 2008/2008). Mayer J.D et al. (Emotional Development and Emotional Intelligence: Educational Implications, 1997/1997, pp. 3—34) stated that on one side of the continuum, people have a greater ability to process the emotion-relevant stimuli effectively, which is associated with an array of benefits. On the other hand, most people have difficulties interpreting their own emotions, managing emotional

Emotional AI and Human-AI Interactions in Social Networking
ISBN: 978-0-443-19096-4
https://doi.org/10.1016/B978-0-443-19096-4.00007-9

outbursts; to handling life pressures. This in turn will harm the self and also the society's growth. The need for emotional quotient (EQ) has created a thirst for all the professionals and people have to work toward building an emotional well-being. Goleman (1998/1998) has developed the elements of emotional intelligence that are

- Self-awareness
- Social awareness
- Self-management
- Relationship management

Nowadays, it has become more obvious how emotionally distant people are from one another. Many technical developments throughout the years have reduced our need for social skills, which inadvertently disguised the long-term decline in human interaction. Understanding and dealing with emotions is an integral part of life. Most specifically, in an educational setup, it is the significant component that helps the instructors to get connected with students.

2. Emotional intelligence

Emotional intelligence aids the teacher in gaining deeper understanding and insight of the learner. Teachers are in the highest need to know the needs of the students in order to set individual specified goals. AEI assists in recognizing and addressing emotional states.

2.1 Theories of emotional intelligence

Many models that try to conceptualize and gauge this construct have been developed as a result of research on emotional intelligence. A variety of ideas have been developed as a result of the concept of emotional intelligence garnering more and more attention over time.

2.1.1 Ability model

The ability model, first put forth by Mayer and Salovey, is one of the best—known theories of emotional intelligence. According to this view, emotional intelligence refers to a set of mental abilities that people may use to recognize, understand, and manage their emotions. The concept distinguishes four domains of emotional intelligence: emotional perception, emotional use as a tool for thought, emotional understanding, and emotional management (Emotional Development and Emotional Intelligence: Educational Implications, 1997/1997, pp. 3—34).

2.1.2 Trait model

The trait model of emotional intelligence is an additional theory that was put forth by Petrides and associates in 2004. In accordance with this theory, emotional intelligence is a group of personality traits that include emotional self-awareness, emotion management, and empathy. The approach classifies emotional intelligence into three categories: sociability, self-control, and well-being (Petrides & Furnham, 2006).

2.1.3 The emotional intelligence mixed framework

Daniel Goleman created the mixed model of emotional intelligence in 1995, which combines aspects of the ability and trait models. The competencies related to self-awareness, self-regulation, motivation, empathy, and social skills are referred to in this approach as emotional intelligence. The model breaks down emotional intelligence into five different subcategories: self-awareness, self-regulation, motivation, empathy, and social skills (Goleman, 1995/1995).

2.1.4 The Bar-On model of emotional intelligence

Reuven Bar-On in the year 1997 formulated of the Bar-On model of emotional intelligence refers to emotional intelligence as a set of emotional and social competencies, including flexibility, general mood, interpersonal, intrapersonal, and stress management. The model breaks down emotional intelligence into 15 subscales, including emotional self-awareness, empathy, social skills, stress tolerance, and impulse control (Bar-On, 1997/1997).

2.1.5 The genos model of EI

Dr. Ben Palmer of Swinburne University created the Genos model of emotional intelligence in 2001. It is described as a collection of competencies and emotional intelligence that allows individuals to skillfully notice, communicate, comprehend, and regulate their own and other people's emotions. The emotional self-awareness, emotional expression, emotional reasoning, emotional regulation, emotional management of others, and emotional self-management are the six basic emotional intelligence skills identified by the model (Palmer et al., 2001).

These models have greatly aided in the development of various training and evaluation programs as well as our knowledge of emotional intelligence. While there is some overlap between these models, it is crucial to remember that each one offers a different viewpoint on emotional intelligence and has advantages and disadvantages.

Generally speaking, the study of emotional intelligence has significant ramifications for a variety of industries, including psychology, education, business, and healthcare. A person's relationships, wellbeing, and ability to succeed in both their personal and professional lives can all be improved with an understanding of emotional intelligence.

3. Artificial intelligence

Artificial emotional intelligence is the term used to describe a machine's capacity to identify, comprehend, and react to human emotions. To replicate emotional intelligence similar to that of humans, it integrates machine learning, natural language processing, and other cutting-edge technology. Using body language, facial expressions, voice tonality, and other nonverbal clues, AEI systems can decipher human emotions. AI enables robots to imitate human behavior, learn from their errors, and adapt to new inputs. Computers may be made to carry out particular tasks with the aid of technologies by digesting enormous amounts of data and seeing patterns in the data. The advancement of AI has been depicted in movies and various sources signaling the benefits of AI. AI has developed to offer a wide range of specialized advantages in every field.

Artificial intelligence encompasses two types of learning which actually make the machines perform human-like activities.

- Machine learning: The subject matter being studied "machine learning" makes a machine able to educate itself instead of being specifically coded. Through the use of machine learning, computers can grasp structure and learn from experience. We now use machine learning in a wide range of industries, and as it develops, it provides us with security, stability, and increased dependability.
- Deep learning: Another technique to advance the technological side of machine learning is through deep learning. Object identification, visual object recognition, sophisticated speech recognition, and blooming in the processing of pictures, movies, and audio are a few examples. Deep learning offers the major benefit of improving performance more quickly the more compressed the data are (LeCun et al., 2015/2015).

Though having a lot of benefits, AI also has its drawbacks such as frequently monitoring the devices, customizing, and updating knowledge to the device whenever needed.

3.1 Importance of artificial intelligence

Fong, T., and Nourbakhsh, I (Fong & Nourbakhsh, 2013/2013) stated that there are numerous important AEI applications in various sectors. For instance, in the healthcare industry, AEI can assist clinicians in better comprehending the emotional condition of their clients, leading to better care. The creation of instructional robots that can perceive and react to students' emotions in the educational setting would benefit their learning experience (Gupta & Gedeon, 2017/2017). Additionally, the application of AEI in customer care might increase client satisfaction by way of tailored exchanges (David & Deangelo, 2018/2018).

3.2 Implementation of artificial intelligence

The application of AEI involves the fusion of numerous technologies, including computer vision, machine learning, and processing of natural language. For AEI systems to learn and develop their emotional identification abilities, a tremendous amount of data is necessary. These data are analyzed using machine learning techniques, which are then utilized to create emotional recognition models. Moreover, nonverbal clues like facial gestures and voice tonality are decoded using computer vision and voice recognition technology (Gunes & Schuller, 2013).

4. Emotional artificial intelligence

Affective computing is the field that gave rise to the concept of studying human emotions in general. There are multiple devices operating in a way to recognize, analyze, and respond to emotional patterns. With the development of technology, it has become clear that comprehension of both the emotional and cognitive aspects of human interaction is vital. Another kind of machine learning called deep learning trains a computer to carry out activities similar to those carried out by a person. Instances encompass forecasting outcomes and detecting speech patterns in the cognitive aspect (LeCun et al., 2015/2015). The concept of artificial EI revolves around the same pattern as cognitive AI. AI would never have evolved as much as it is today without deep learning.

Stuart (1995/1995) stated that AI-equipped machine's core purpose includes learning, investigating, analyzing problems, coming up with reliable solutions, performing predictive analysis on the subject, and improving hazard identification for specific features.

The technological advancements have led the devices or machines to recognize the individual's emotional state. Making a machine, computer, or software system think and act like a human is known as AI. The AI should be logical and possess human-like innate intelligence. AI that possesses emotional intelligence can more closely resemble humans, which can help it better understand society from the standpoint of an individual. Emotional AI aids in various firms. Several researchers have begun to develop machines that can elicit, sense, manage, understand, and express emotions (Sood, 2008).

People express their emotions through a variety of nonverbal indicators, including gesture, mannerisms, voice modulation, and facial expressions. So, AEI should be able to perceive emotions through a variety of pathways, just like humans do.

The classification of artificial emotion software systems by Rumbell (2012/2012) was based on four queries: The meaning of action selection method, structure of action selection method, possession of emotional roles, and emotional model employed.

4.1 Challenges in implementing emotional AI

The complexity of human emotions, the absence of standardized emotional datasets, and the ethical issues surrounding the use of AEI are only a few of the major difficulties facing the development of AEI systems. The difficulty of preventing bias in AEI systems' emotional recognition is another (Calvo & Peters, 2014/2014). But there is still much work to be done. In summary, AEI is an exciting area with enormous potential in a variety of industries. Yet there is still a lot of effort to be done to overcome the obstacles and create trustworthy AEI systems that can correctly identify and react to human emotions.

4.2 Artificial intelligence in educational sector

Advancement in artificial intelligence improves learning and educational experience. It is considered one of the newest technologies (Ross & Issroff, 2018/2018). The entire "learning" experience" will change when AI is applied in education. There are just a few things that drove transformations in the global education industry to shift the system (Blaylock, 2019/2019). That are:

- Adjustable learning environment
- New opportunities

- Improved efficiency
- AI-driven education platform

With such advancements in education that improve learning, AI may be considered one of the newest technologies.

The objective of an entity designing an AI application is in education. Task of capturing large volumes of curricular resources with metadata relating to the subject area, pertinent grades, language, type of resource, file format, level of resource, and role (introducing a topic, reinforcing learning) in teaching. Secondly, the actual use of these resources in various learning contexts, the teacher's pedagogical activities, and the responses of the students to the content would be another task. This is typically done through assessment procedures that aim to determine the student's level of learning and understanding before and after the interaction with a resource (Gurumurthy & Yogesh, 2019/2019). AI enhances the learning experience in many ways that support the curriculum and mental health of the learner and teacher. The employment of AI in education is fraught with difficulties. Some of these apply to the use of AI in various industries. The first is that the AI's algorithms are not unbiased or objective, presuming that such a thing even exists. In the code produced by machines, programmers' prejudices are exacerbated. This can occasionally be seen in the results, such as when criminal profiling algorithms are used to African-Americans in the USA with greater (and unfairly harsher) severity, as well as in other contexts like credit score and insurance premium calculation (Gurumurthy & Yogesh, 2019/2019).

4.3 Emotional artificial intelligence in educational sector

The deficiencies of the education sector in particular have come to light; never before has the distinction between effective educators and less effective educators been so clear. Understanding someone's feelings requires being able to recognize, sort through, and interpret both nonverbal and verbal signals. At the initial stage the emotional AI has been used in students learning in the role of grading and evaluations which supports the educators. When applied to children, emotional AI raises questions about how child behavior and subjectivity will be transformed into profit Lupton and Williamson (2017).

Emotional AI technology can be applied across a wide range of sectors. By making major investments in technology, the applications of technology across diverse industries are extended. Considerations relating to child bodies and profiling are part of a larger conversation about children and media, internet screen time, and the growing trend for children to be "always be on" in all aspects of childhood (such play, interaction, education, and health enhancement Livingstone et al. (2018).There are few applications of emotional AI, which are added as follows, to monitor students' emotional states during courses, sensors like microphones and camcorders can be employed. Emotional AI can assess whether students are pleased or unhappy with the teachings based on whether a task is either too challenging or too easy. As a result, instructors can vary their approaches to adapt the class load. A similar approach may be used to test learning software models for online learning. Another use in education is to help autistic persons recognize other people's emotions in a classroom setting. Emotional AI aids autistic children in supporting the identification of nonverbal cues. School psychologists or professionals could greatly facilitate their children by using emotional AI during counseling sessions to better track and comprehend children's emotional problems.

Data have become a capital in today's commercial and technological world and is crucial to the success of the company. Without a doubt, "Big Data" is a major factor in the development of artificial intelligence and advantages in artificial emotionality. According to Gartner, "big data" encompasses data resources with massive density, speed, and diversity that call for creative, expensive methods of information processing to improve understanding and choice. Advanced learning algorithms that can record, evaluate, and retain human activity such as wants, emotional states, prejudices, and actions based on communication, friendships, and cultural context are being designed in response to the growth and adoption of big data. This is how equipment and technologies for artificial intelligence and AEI are being developed, using the vast amounts of data and information that technology companies like Facebook, Google, Microsoft, Twitter, and others have collected from billions of people.

One must first comprehend, at the very least at a fundamental level, what emotional AI encompasses in order to comprehend the role that emotional AI can play in the education sector. It is possible to divide emotional AI into two parts.

- Establishing a database of emotional reactions through emotion recognition
- Using knowledge gained through emotion recognition is known as an emotional response.

Though we give off a lot of cues about our emotions, the easiest things for the education sector to focus on are facial expressions, sounds, and gestures.

4.4 Devices of emotional artificial intelligence to support learners and students

Although emotion-sensing technologies are still in its inception, their business functions are now widely used in our day-to-day lives, blossoming into a USD 21.6 billion sector that is predicted to grow by over twofold in value by 2024 Crawford (2021).

Basically, the way that these programs work is that systems receive signals of behavior related to feelings, predicted thoughts, actions picked up in speech, thoughts, and beliefs. For instance, if an individual's eyeball dilates, it then gives signals about the physiological changes. In addition to looking for means of improving automation better at what it does now, developers are also investigating methods for making automation relate to how the population thinks about their life. This is where the perspective of emotional artificial intelligence is relevant. The majority of the time, voice assistants are taught to answer frequent questions. Aspects like irritation, amusement, frustration, or sarcasm could go unnoticed. This problem is resolved by emotional AI, which recognizes and understands emotional measurements and intonation in a specific voice to determine the significance of the conversation.

Smartphone applications in the realm of education, such as Class Dojo, give teachers psychometric profiles of their students, enabling them to grade and recognize positive conduct while imparting a lesson Williamson (2021). Meanwhile, the interactive toy Moxie promotes a child's emotional, social, and psychological growth Lyles (2020).

Dutch bank and Philips collaborated to produce a "rationalizer" bracelet. This tool is intended to assist the trader by preventing him from making irrational decisions. It checks the wearer's pulse rate to determine whether they are in an emotionally charged condition and alerts them to quit making irrational decisions. Google's Brain Power smart glasses are a device that sees and hears unique feedback tailored to the user's emotional state based on facial expression. It can aid those with autism in comprehending emotions better.

4.5 Face analysis

There are countless possible facial expressions that can be recognized by facial gestures and it is universally applicable. Other characteristics, such as whether or not a person is happy, how expressive a face is, etc., are now being incorporated into machine learning models used for emotional analysis. They determine how someone feels by giving them a blend of these emotions, rather than necessarily categorizing them into one of these strong emotional categories (angry, pleased, and surprised).

This makes it possible to reasonably estimate how someone is feeling at any particular time. But facial analysis still has certain limitations and cannot fully convey a person's emotions on its own. Speech and gesture analysis can help.

4.6 Voice analysis

The conventional method for analyzing human speech makes use of a variety of metrics, each of which is handled as a distinct "emotional measure." These metrics include, for instance: voice affectivity, engagement of the speaker, and control of the speaker over situations.

Automatic audio/speech detection schemes analyze the emotions from an audio clip. The human voice and emotion detection is a crucial subject research area in the present time. The speech emotion detection system is categorized into three stages: Extraction of features, reduced feature dimension, and engine for classification.

4.7 Gesture analysis

In order to effectively express one's own feelings and comprehend those of others, gestures appear to be quite important. Because of social distancing, most meetings and courses take place online, which restricts our capacity to use gestures as a way of communication.

5. Ethics of EAI in educational sector

A high level of awareness of data security and Internet privacy is becoming more prevalent. Educational institutions employing emotional AI must be transparent and open about their digital ethics in order for emotional AI to function properly and not have the reverse impact, which is to frighten people rather than give them positive rewards.

The educational institution should be clear and precise for detecting any specific emotional patterns and the purpose of collecting data should be predefined. For example, the educator can work on the objective of why emotional AI.

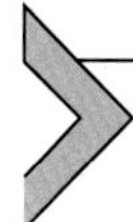

6. Future of emotional artificial intelligence in education

In the presence of the UNESCO Director General, the UNESCO MGIEP brings together experts, academicians, researchers, and members of civil society to develop a strategy for incorporating AI for social and emotional learning in the 21st century while keeping in mind the legal and practical considerations surrounding such an incorporation. Here, the main difficulty is in reimagining educational systems to move beyond simply implementing AI to support current pedagogical practices of the transmission-model of education. Concerns have also been raised concerning the hazards and drawbacks of AI in education.

7. Conclusion

The complex modulators that cause affect and emotional states need to be better understood by the scientific community. Perhaps, expanding global alliances and collaborations can remedy this predicament and aid in the creation of more sophisticated technology that can better handle factors like gender, ethnicity, attitudinal diversity, and cross-cultural emotional complexity. AEI assists in recognizing and addressing emotional states. Affective computing field entirely deals in developing devices that are able to perceive, comprehend, understand, and imitate human emotions. Devices can listen to voices, gaze through facial expressions, outlook the gestures to reflect on emotions in a way of augmented human. Although emotional AI cannot completely replace teachers, it may aid them in certain circumstances. Teachers might be able to maximize each student's learning potential with the use of emotional AI. Global technology uses artificial intelligence to address each child's individual emotional needs in addition to enhancing pupils' academic performance while receiving homeschooling.

References

Bar-On, R. (1997). *Bar-on emotional quotient inventory (EQ-i): Technical manual. Multi-health systems* (Original work published 1997).

Blaylock, Jeff. (2019). *The top five changes that occur with AI in Education* (Original work published 2019).

Calvo, R. A., & Peters, D. (2014). Promoting EMMA: Addressing the challenges of developing emotional artificial intelligence. *Emotion Review, 6*(4), 319–329 (Original work published 2014).

Colman, A. (2008). *A dictionary of psychology (Vol. 9780199534067)*. Oxford University Press (Original work published 2008).

Crawford, K. (2021). *Artificial intelligence is misreading human emotion*. Retrieved from https://www.theatlantic.com/technology/archive/2021/04/artificial-intelligence-misreading-human-emotion/618696/.

David, D., & Deangelo, L. (2018). *Emotion AI and the future of customer service* (Original work published 2018).

Emotional development and emotional intelligence: Educational implications. (1997). Basic Books. (Original work published 1997).

Fong, T., & Nourbakhsh, I. (2013). Emotional intelligence in healthcare robots. *International Journal of Social Robotics, 5*(4), 523–530 (Original work published 2013).

Goleman, D. (1995). *Emotional intelligence: Why it can matter more than IQ*. Bantam Books (Original work published 1995).

Goleman, D. (1998). *Working with emotional intelligence*. Bantam Books (Original work published 1998).

Gunes, Hatice, & Schuller, Björn (2013). Categorical and dimensional affect analysis in continuous input: Current trends and future directions. *Image and Vision Computing, 31*(2), 120–136. https://doi.org/10.1016/j.imavis.2012.06.016

Gupta, R., & Gedeon, T. (2017). Affective robots in education: Systematic review and future challenges. *Journal of Educational Technology & Society, 20*(1), 136–148 (Original work published 2017).

Gurumurthy, K., & Yogesh, K. (2019). *Exploring AI in Indian school education* (Original work published 2019).

LeCun, Y., Bengio, Y., Hinton, G., Livingstone, S., & Mascheroni, G. (2015). European research on children's internet use: Assessing the past and anticipating the future. *New Media and Society, 20*(3), 1103–1122 (Original work published 2015).

Livingstone, S., Mascheroni, G., & Staksrud, E. (2018). European research on children's internet use: Assessing the past and anticipating the future. *New Media and Society, 20*(3), 1103–1122.

Lupton, D., & Williamson, B. (2017). The datafied child: The dataveillance of children and implications for their rights. *New Media and Society, 19*(5), 780–794.

Lyles, T. (2020). *Moxie is a $1,500 robot for kids*. Retrieved 2021 September 08 from https://www.theverge.com/2020/5/13/21257821/moxie-robot-kids-educational-tech-embodied-price.

Palmer, B., Walls, M., Burgess, Z., & Stough, C. (2001). Emotional intelligence and effective leadership. *The Leadership & Organization Development Journal, 22*(1), 5–10. https://doi.org/10.1108/01437730110380174

Petrides, K. V., & Furnham, A. (2006). The role of trait emotional intelligence in a gender-specific model of organizational Variables1. *Journal of Applied Social Psychology, 36*(2), 552–569. https://doi.org/10.1111/j.0021-9029.2006.00019.x

Ross, L., & Issroff, K. (2018). Future of education and skills 2030: Conceptual learning framework. Education and AI: Preparing for the future and AI, Attitudes and values.

In *Eighth informal working group (IWG) meeting* (pp. 29–31). Conference Centre (Original work published 2018).

Rumbell, T., Barnden, J., & Denham, S. (2012). Wennekers, emotions in autonomous agents: Comparative analysis of mechanisms and functions. *Autonomous Agents and Multi-Agent Systems, 25*. https://doi.org/10.1007/s10458-011-9166-5 (Original work published 2012).

Sood, S. O. (2008). Emotional computation in artificial intelligence education. *AAAI Workshop - Technical Report, WS-08–02*, 74–78.

Stuart, J. R., & Norvig, P. (1995). *Artificial intelligence a modern approach* (Original work published 1995).

Wang, Li (2022). Exploring the relationship among teacher emotional intelligence, work engagement, teacher self-efficacy, and student academic achievement: A moderated mediation model. *Frontiers in Psychology, 12*. https://doi.org/10.3389/fpsyg.2021.810559

Williamson, B. (2021). Psychodata: Disassembling the psychological, economic, and statistical infrastructure of "social-emotional learning". *Journal of Education Policy, 36*, 129–154.

CHAPTER SIX

Emotional AI: Neuroethics and Socially aligned networks

Markus Krebsz[1,2], Divya Dwivedi[2,3]
[1]University of Stirling, Management School, Stirling, United Kingdom
[2]Woxsen University, Decision Sciences & Artificial Intelligence, Hyderabad, Telangana, India
[3]Supreme Court of India, New Delhi, India

GLOSSARY

Artificial Intelligence (AI) system defined by the OECD (OECD - Recommendation of the Council on Artificial Intelligence, 2019) as "an AI system is a machine-based system that can, for a given set of human-defined objectives, make predictions, recommendations, or decisions influencing real or virtual environments. AI systems are designed to operate with varying levels of autonomy."

Emotional AI is one of the more recently emerging domains of artificial intelligence (AI), which is also referred to as emotion AI or affective computing. It largely covers the ability of computational systems and machines to study, analyze, and interpret humans via mostly nonverbal features such as gestures, facial expressions, body language, as well as other factors including human voice to establish and determine humans' emotional states. Naturally, this is one of the more controversial areas of AI and consequently subject to greater scrutiny by AI Ethicists.

Neuroethics An interdisciplinary field devoted to the study of the ethical, legal, policy, and social implications of advances in neuroscience and their impact on people and society.

Neuroprivacy A neuroethical concept of privacy concerns pertaining to neural information that is obtained through imaging or diagnostic technologies and the use of the information in legal and societal contexts.

Pragmatic neuroethics A practical, solution-oriented approach to neuroethical inquiry that privileges empirical analyses over a priori moral principles and emphasizes real world circumstances, pluralism, and multidirectional, inclusive deliberations.

Social network An online service or site through which people create and maintain interpersonal relationships.

Socially aligned network An online service or site or gaming suite or virtual-/augmented-/extended-reality (VR/AR/XR) or metaverse or similar technological ecosystem through which people communicate, create, date, compete, and/or challenge each other. In addition, they use socially aligned network(s) also to establish and maintain relationships that are based on either their real or fabricated online identities and their aligned or mis-aligned common interests. Although they may overlap with the more traditional social networks, membership and activities are more characterized by an alignment of participants' common interests and less focus on interpersonal rapport.

Emotional AI and Human-AI Interactions in Social Networking
ISBN: 978-0-443-19096-4
https://doi.org/10.1016/B978-0-443-19096-4.00002-X

1. Introduction

While the preceding book chapters within this section explore specific applications of human—AI within the social network context, this chapter looks at applicable ethical studies from a more universal angle. It further aims to identify a suitable ethics principles baseline. It then considers current advances in neuroethics and reflects upon transferability of some of the ethical principles to emotional AI and human—AI interactions across social networking and beyond.

Given emotional AI, often referred to also as affective computing, is a relatively new area of research with some of it still representing uncharted territory, and under ongoing scrutiny by AI ethicists, the following definitions will help setting the scene while it is important to acknowledge that, due to the dynamic nature of this field, they are not set in stone. Where our own views go beyond this, we will specifically highlight relevant deviations from existing taxonomy.

1.1 Setting the scene

1.1.1 Social networks

A **social network** can be largely defined as an online service or site through which people create and maintain interpersonal relationships. This widely accepted term is closely aligned with the definition and data provided by Dixon (2023a,b), (see Fig. 6.1) that is based on *Internet users who use a social network site via any device at least once per month*. A recent visualization by Routley (2022) of the Worlds' top social media and messaging apps highlighted both clustering of apps by some of the Big Tech players such as Meta, Tencent, ByteDance, Microsoft, and Amazon as well as the huge amount of monthly average users for each of those platforms in comparison to the World population according to UNFPA (2022).

Although an overlap of users of these social media networks is observable, likely caused by users often signing up with a multitude of these services at the same time, it is also obvious that a large percentage of the global world population frequently interacts with these social networks.

Consequently, emotional AI used in these ecosystems to influence the human—AI interactions poses a global risk to freedom of thought, and hence, we have taken a closer look at both the existing AI ethical principles baseline as well as well any additional advances in the ethics research branches that may need deploying within the social network context.

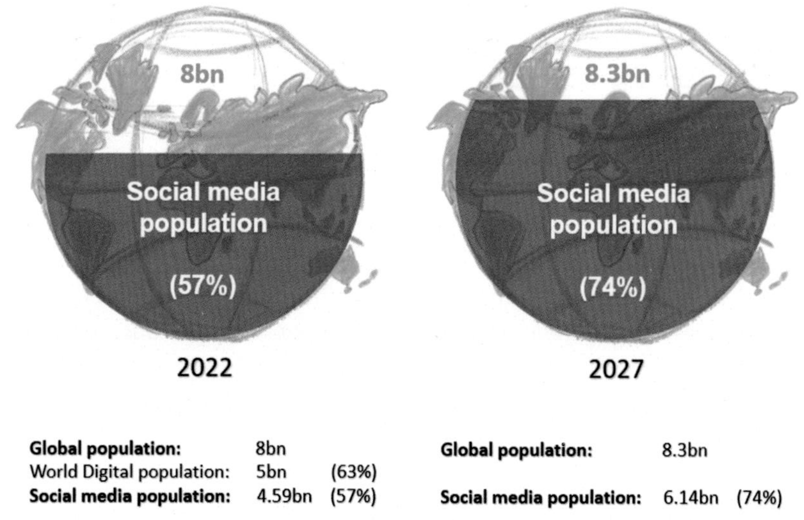

Figure 6.1 Global social media penetration from 2022–27. *Original drawing of the world globe, 2020, Markus Krebsz, All rights reserved.*

1.1.2 Socially aligned networks

In contrast to the pure social media network and messaging apps landscape exhibited above, our research goes beyond this as we also consider the many in-gaming communities (e.g., PlayStation, Xbox, Steam networks), dating sites (e.g., Badoo, Tinder, Plenty of Fish, Grindr, Bumble, Adult Friend Finder, etc.) as well as newly emerging metaverses and virtual-/augmented-/extended-reality (VR/AR/XR) community networks within the scope of this study.

In order to distinguish traditional social networks within the context of our own studies, we have coined the term **socially aligned network** meaning an online service or site or gaming suite or virtual-/augmented-/extended-reality space or metaverse or similar technological ecosystem through which people communicate, create, compete, and/or challenge each other. In addition, these people also use such socially aligned networks to establish and maintain relationships that are based on either their real or fabricated online identities and any aligned common interests.

Although they may overlap with the more traditional social networks, membership and activities are more characterized by an alignment of participants' common interests and relatively less defined by interpersonal rapport and social cohesion. Despite our definition being considerably wider than the contemporary view and in light of a multitude of existing research analyzing the mechanics and ethics of purely human-to-human interaction within social networks, our ethical studies presented within the context of this book exclusively investigate the human-to-AI (individual persons to AI machine) interface.

1.2 Human component and bias

When people make decisions, they typically let their personal feelings or opinions influence their choices. This is called bias. AI is a computer system that can be programmed to make decisions for people and on their behalf. Because AI is not influenced by personal feelings or opinions, it can—in theory—be more objective or fair. However, we must remember that humans design and deploy AI thereby embedding their views into the systems they build.

Consequently—in practice—systems built by humans will always intrinsically be somewhat biased, and hence it is important to check that the AI system is working as it should and system-specific bias is studied, understood, documented and continuously monitored throughout the AI system's lifecycle. We are of the view that bias cannot be eliminated, however, it is possible to mitigate some of its undesired effects and manage any residual risk within AI systems accordingly.

1.2.1 AI propagates and potentially amplifies human biases

Unlike humans, an algorithm is a set of rules or instructions that cannot dissent, disobey, or make exceptions. If data, that are fed into the systems, contain intrinsic bias (which it often does), there will be bias in the output data too. For example, if you were to ask a group of people their favorite color and 80% said blue, then the algorithm would output that blue is the favorite color. But unfortunately, people's obsession with predicting future outcomes results in the development of technology that re-enacts the past and, in doing so, perpetuates (and potentially amplifies) its mistakes in future decision-making.

1.2.2 Differences in definitions, axioms, and thematic domains

Furthermore, a significant challenge for AI is how to account for all the data's historical, social, and racial attributes when producing judgments using algorithms. AI can only be considered as good as the minds of those who design it, that is, the humans (Lin, 2021). AI does not appear to be a common identifier across the board. Axioms about AI differ, as do the definitions regarding the thematic domains themselves. For instance, a lawyer and a statistician might have completely different views of what represents a fair outcome (Ethics Washing Is When Ethics Is A Substitute for Regulation, 2018).

1.3 The AI definition and ethics baseline

Although a wide range of AI definitions exist, we use the OECD's Recommendation of the Council on Artificial Intelligence nomenclature, which explains that "*an AI system is a machine-based system that can, for a given set of human-defined objectives, make predictions, recommendations, or decisions influencing real or virtual environments. AI systems are designed to operate with varying levels of autonomy.*" (OECD - Recommendation of the Council on Artificial Intelligence, 2019) as our baseline AI definition for this chapter.

Given that the concept of AI first surfaced over 70 years ago, complementary ethical studies have reached a natural level of maturity, and consequently, we have been taking the global ethical AI landscape into account when trying to establish an AI ethics principles baseline. In Fjeld et al. (2020) published an article, white paper and visualization following their attempt to map consensus in ethical and human rights-based approaches to principles for AI highlighting that "*there has been little scholarly focus on understanding these efforts either individually or as contextualised within an expanding universe of principles with discernible trends.*" Their study compared the content of 36 documents relating to AI principles and yielded a data visualization, as a result of analyzing those. Importantly, their work identified the following eight ethical key thematic trends.

1. Privacy,
2. Accountability,
3. Safety and security,
4. Transparency and explainability,
5. Fairness and nondiscrimination,
6. Human control of technology,
7. Professional responsibility, and
8. Promotion of human values.

Further, and maybe more importantly, their study suggests that these universal eight themes may represent what they call the "normative core" (Fjeld et al., 2020) of AI ethics and governance principles.

As part of our research, we have globally seen consistent references to these eight key thematic trends, and hence, we consider those as starting point and ethical baseline for our own analysis, while at the same time being mindful of their (Fjeld et al., 2020) study cautioning of the contextual setting including cultural, linguistic, geographic, organizational, governance, regulatory, and relevancy implications. The current rules for AI ethics do not constitute any sort of regulatory forces or legislative requirements. Because of this, many norms are seen as merely pointers for ethical governance, while some have been perceived as means of public appeasement or an avoidance of tough choices and have encouraged extensive ethical consumerism (Kerrigan, 2022).

1.3.1 Traditional ethics limitations concerning emotional AI

Traditional AI ethics approaches alone may not be sufficient to address the additional risks that human—AI and emotional AI pose when aimed at directly influencing humans' experiences, emotions, thinking, decision-making, and behaviors. Consequently, where the human—AI interface within a socially aligned network is concerned, additional ethics methods will need to be deployed to bridge the gaps where emotional AI is concerned.

Although this book's title suggests its focus on social "advances" in neuroscience, an exploration of emotional AI and human—AI interactions across social networks will also need to carefully consider risks that naturally come when neurotechnological innovation undoubtedly impacts both individuals' thinking, emotions, and behaviors as well as humanity's collective responses and actions.

Individuals, how they think, respond, and interact on **socially aligned networks** are hugely profitable for Big-Tech companies who have naturally access to all of these data and then deploy AI for behavioral pattern analysis to derive value from the intelligence and insights they have gained by deploying algorithms in this fashion.

As Alegre (2022) in "Freedom to Think" suggests, *"how we think and how we can be made to think and behave is what is commercially valuable. Because if you can understand and control how we think and feel, you can control what we buy, what we do and how we vote."*

Taking this increased use of emotional AI and the direct impact it has on humanity's mental well-being in account, it is our view that emotional AI in general and human—AI interaction within socially aligned networks specifically, needs to look closely at ways and means to manage risks that come with such uses. As a result of these observations, we have been looking toward and subsequently been borrowing approaches from two scientific fields where both giving explicit consent and human brain—interface interactions are typically exposed to specific ethical considerations: medicine and neurotechnology.

1.3.2 Informed consent for emotional AI

While it may not always be that obvious to individual users, social media platforms, in general, draw them in by offering incentives for a particular (targeted) service aimed at selling tailored product(s). The more platform users are directed to and matched with appealing products and services, the more the platform can expect to realize financial profits in return.

To learn about user preferences, social media companies use behavioral profiling and nudging to get them to arrive at certain choices. Can society at large apply the same policies to everything it does (Custers & Fosch-Villaronga, 2022)?

Conventional governance is important in providing a fair medical science system to patients who have consented to allow their brain data to be examined. Informed consent in the neuroscience field could emphasize the autonomy and privacy of patients by making sure they know how the information they provide is bound and protected (Ahluwalia, 2021).

So, while informed consent is a standard and normative term in the medical world including neuroscience, it appears to be amiss in all of the socially aligned networks, which made us consider further whether a similar informed consent should apply to emotional AI and which additional safeguards, if any, may be required for users exposed to human—AI.

1.3.3 Emotional AI ethics: Borrowing from neuroscience

Given the close relationship of socially aligned networks and the proximity to the human—brain interface, one relatively new area of ethical studies and research is within the neuroscience field itself. Accordingly, we have consulted with and are grateful to Dr. Moritz J. Maier, project lead at the Fraunhofer Center for Responsible Research and Innovation (CeRRI) of a multiyear study aimed at the participative development of a Code of Conduct for the EU in the area of noninvasive brain stimulation, who kindly shared with us ethical observations and papers from the project's comprehensive literature review collection. The sections in this chapter covering neuroethics are largely inspired by and based on research and insights by Dr. Moritz J. Maier.

1.4 Current scope and predicted growth of the world digital population

The gravity of ethical issues where socially aligned networks is concerned becomes obvious when considering both current and predicted social media penetration rates: With three quarters of the world population forecast to be active on social media by 2027 (see Fig. 6.1), the AI tools deployed by Big Tech companies that provide those ecosystems are both hugely powerful and, without any appropriate external regulatory oversight and robust internal governance procedures, can become detrimental to a large proportion of humanity.

As highlighted in Fig. 6.1, based on data sourced from Statista (Dixon, 2023a,b) and the United Nations Population Fund (UNFPA, 2022) [4], the global social media population in 2022 represented around 57% and is expected to grow to almost three quarters by 2027 according to Dixon (2023a,b). We further hypothesize that the global socially aligned network population will likely be somewhat even greater than the estimated 57% in 2022 and 74% in 2027, respectively.

We assume global socially aligned networks represented 60%–65% of the world's population in 2022 and estimate, somewhat conservatively—with the inclusion of dating sites and the relatively recent emergence of new metaverses, virtual- and augmented-reality and crypto-/NFT-based gaming ecosystems—the penetration rate for socially aligned networks to increase globally to 80%–85% by 2027.

1.4.1 Digital exclusion

While this particular section of the book specifically considers human–AI in the context of social networks, it is worthwhile highlighting that in 2022, around 27% of the world's population was not digitally connected and, as a result, excluded from any digital activities, including both social media and socially aligned networks. Although it is expected that this proportion impacted by such digital divide will reduce over the next decade, an observable digital exclusion will persist and will likely increasingly concentrate in and penetrate impoverished, illiterate parts of the world.

1.4.2 Possible reasons for staying analogue

Further, the difference between the world digital population (63%) versus social media population (57%) suggests that there is currently a notable absence of some of the world's digitally connected citizens from both social media sites and socially aligned networks. Due to the lack of data highlighting specific reasons for this observable absence, we hypothesize that this is likely attributable to a mix of deliberate personal choice to abstain, government bans of social networks in some geographies mixed with a range of other reasons, including cultural preferences.

Now that we have a grasp of the size of socially aligned networks within the emotional AI context, we are providing you with an overview of the methods chosen for our research and our thinking behind those. At the same time, we would like to remind the readers that this is a rapidly evolving area of great dynamism and as such are not claiming this is the best or only way of addressing the ethical challenges posed by socially aligned networks and we are keen on hearing from how our readers would address these emerging ethical challenges.

2. Research methods

The starting point for our study's main focus was three-fold with our specific research focus aimed at.

1. Quantifying humanity's breakdown by digital activity and penetration of socially aligned networks,
2. Qualitatively exploring the availability of existing ethical key thematic trends of AI principles and determination, identification of additional contextually necessary AI principles, and highlighting observed gaps in contemporary ethical principles for the human—AI interface and emotional AI, and
3. Specifically excluding human-to-human interactions within a social media and socially aligned network context. Such interactions and related human behavioral traits are already well-researched and, hence, have been out of scope for the purpose of this specific study.

2.1 Quantifying socially aligned network users

We took data from Statista decomposing the global world population into population cohorts for World Digital and Social Media population in 2022. Where available, this was compared with the 2027 predicted Global World population (UNPFA, 2022) and the corresponding social media population cohort.

When dissecting the underlying data further, we hypothesized that the actual number of global social media users would likely be higher when accounting for other technological ecosystems.

Naturally, there will be an overlap from users of those ecosystems with the more traditional social media networks (e.g., Meta, Twitter, Instagram, TikTok, etc.) while these newly emerging ecosystems continue to grow. In order to reflect this and distinguish different population cohorts for the purpose of this study, we coined the term "socially aligned networks," which we further define in Section 1.1.2.

2.1.1 Mapping the ethical studies landscape

In order to establish an ethical studies baseline, we looked toward other comparative studies focused on ethical AI and governance principles. Fjeld et al.'s (2020) curated, reviewed, hand-coded mapping and visualization exercise was deemed suitable given the nature and limitations involved (and, as described in detail in their work). In light of observing very similar thematic trends globally (albeit with researchers sometimes using different nomenclature), we concluded that the eight key thematic trends identified covering privacy, accountability, safety and security, transparency and explainability, fairness and nondiscrimination, human control of technology, professional responsibility, and promotion of human values represent a sound ethical studies baseline of AI ethics and governance principles for our own study.

In order to cater to ethical challenges of human–AI interfaces deploying emotional AI within a socially aligned network context, we subsequently widened our research and looked toward ethical principles specifically within the neuroscience sector, where AI systems are increasingly becoming directly connected to the human brain—either through noninvasive or invasive means. Finally, informed by several case studies (see next section), both the applicability of the more traditional baseline ethical AI principles in combination with those emerging from the neuroscience field as well as any resulting observable gaps have informed the outcome of our research.

2.1.2 Human–AI interface case studies

Ethical studies of human–AI interfaces within a socially aligned network context require investigating case studies of instances where individuals and/or social groups were subjected to the deployment of emotional AI algorithms, often leading to human disadvantage or detriment. We selected such case studies to consider whether neuroethics alone would be sufficient to address these ethical challenges or if there were any gaps within contemporary thinking and approaches. Given the dynamic nature of this field, we have not always identified answers and invite other researchers to investigate appropriate mitigants further.

2.2 Results

In this section, we discuss our experimental and analytical results.

2.2.1 A rapidly growing digital and connected world

We sourced data from Statista (see Fig. 6.1) decomposing the Global world population (8bn) further into population cohorts for World Digital (5bn or 63%) and social media (4.59bn or 57%) population in 2022. Where available, this was then compared with the 2027 predicted global world population (8.3bn) and the predicted 2027 social media population cohort (6.14bn or 74%). Although the 2022 global social media population represents already a substantial 57% (or 74% in 2027, respectively) proportion, when dissecting the underlying data further, we hypothesized that the actual number of global social media users is likely going to be higher when accounting for other technological ecosystems, such as gaming suites or virtual-reality spaces or virtual-/augmented-/extended-reality (VR/AR/XR) gatherings or metaverses or similar technological ecosystem through which people communicate, create, compete, trade with, and/or challenge each other. Naturally, there will be an overlap from users of the more traditional social media networks (e.g., Meta, Twitter, Instagram, TikTok, etc.) with users of those newly emerging ecosystems. As new ecosystem user numbers continue to grow globally, we expect the global number of users of all of those services to exceed numbers reported under social media population in Fig. 6.1. In order to reflect this and distinguish different population cohorts collectively for the purpose of this study, we have coined the term "socially aligned networks."

2.2.2 The AI ethics principles baseline

The (Fjeld et al., 2020) mapping and visualization exercise established the "normative core" of AI ethics and governance principles by identifying eight key thematic trends observable in most of the 36 curated documents investigated. And, given our study refers to those eight key themes as AI ethics baseline, it is worthwhile further articulating those eight key themes within the context of socially aligned networks that increasingly deploy emotional AI.

1. **Privacy**—the idea behind this principle is that emotional AI systems used for socially aligned networks should respect and protect individuals' privacy both with regards to the use of data for the development of such technology and by providing users with agency over the decisions that is made with their data as well as the data itself and any thought- or emotional state-related inferences.
2. **Accountability**—this principle focuses on the importance of the establishment of emotional AI system governance mechanisms ensuring that accountability for the impacts of those systems within the socially aligned

network is appropriately distributed and that adequate remedies are provided. Such accountability measures may include internal governance bodies such as an ethics committee or an algorithmic risk committee as proposed by ForHumanity and Carrier (2021b).

3. **Safety and security**—these principles underscore requirements for emotional AI systems deployed for socially aligned networks to be safe, performing as intended as well as secure. The security element also means resistance to being compromised by unauthorized parties, covering both cyber and physical intrusion.
4. **Transparency and explainability**—these principles highlight the need for emotional AI systems of socially aligned networks to be organizationally governed and technologically designed and implemented to allow for oversight, including traceability of the data inputs across the whole algorithmic processing and operations lifecycle into intelligible outputs. This also includes full transparency about where, when, and how data inputs and derivations thereof—including biometrically inferred data—are being used as part of the algorithmic decision-making.
5. **Fairness and nondiscrimination**—these two sections of the principles have been receiving most global attention and focus with growing concerns that bias stemming from emotional AI systems within socially aligned networks is already negatively affecting individuals. Both principles aim to maximize algorithmic fairness and promote inclusivity.
6. **Human control of technology**—this principle in particular is gaining increasing traction, and the underlying theme means important algorithmic decisions will ultimately be subject to human review and control - often referred to either as "human-in-the-loop" or "human-on-the-loop."
7. **Professional responsibility**—this principle recognizes individuals' vital involvement in coding and deploying emotional AI systems within a socially aligned network context and requires consultation with diverse inputs and gathering multistakeholder feedback ("DI and MSF") according to ForHumanity and Carrier (2022) ensuring that long-term effects of the algorithmic systems have been addressed in the ecosystem design.
8. **Promotion of human values**—this principle requires that the purpose of the emotional AI system of socially aligned networks are devoted to and the way the ecosystem has been implemented should correspond with core human value principles, promote human objectives and generally elevate humanity's well-being.

We expect to see an increasing emergence of regulatory requirements globally, likely going to be informed to varying degrees by some or all of those ethical AI principles.

2.2.3 Increasing regulatory scrutiny of inferred data considerations

Although our research has been purely focused on the ethical aspect of human–AI interfaces and not on regulatory requirements, we are acutely aware and deeply mindful that all socially aligned networks and the way they use both bulk data as well as inferences from those data points will likely become increasingly subjected to regulatory scrutiny.

A particular area where we expect to see more stringent regulatory requirements is for bulk data sets collected from information inferred from behavioral, physical, and psychological biometric identification techniques as well as other nonverbal communication methods, commonly referred to as "biometrically inferred data (BID)." The biometric identification techniques that are often used to determine such biometrically inferred data highlighted by XR Safety Initiative (n.d.) are.

- Facial recognition,
- Fingerprint verification (Dactyloscopic data),
- Handwritten signature analysis,
- Retinal analysis,
- Iris scanning,
- Voice recognition,
- Recognition of the shape of the Human ear,
- Keystroke analysis,
- Gait analysis, and
- Eye tracking (Gaze analysis).

For example, by tracking eye movements, including duration and width of eye opening and closure, eye status, pupil properties, iris characteristics and facial attributes in conjunction with this, emotional AI deployed can make possible inferences covering a wide range of personal (and often legally protected) attributes including age, gender, geographical origin, cultural background, skills and abilities, personal traits, mental health and workload, as well as constructing and storing the users' biometric identity in its entirety.

Another, closely related to biometrically inferred data, area is the emerging branch of criminology often referred to as "neurocriminology" that deploys algorithms to predict future crimes and tries to determine this on the state of minds of expected criminals. Such approach assumes sanity of the perpetrator and causes potential legal conflicts particularly concerning the so-called "insanity defense" or M'Naghten's rule in the criminal justice system, which assumes that every person is assumed to be sane. Further, in order to establish a defense on the grounds of insanity, it must be clearly proven that, at the time of committing the unlawful act, the accused party was laboring under such a defect of reason from disease of the mind.

Consequently, an emotional AI algorithm may prejudge humans, including their predicted state of mind, and conclude that a crime is likely to be committed on such a basis, meaning it becomes an automated judge, jury, and executioner with a narrow, algorithmically limited and possibly prejudiced view (Dartmouth Undergraduate Journal of Science, 2013). In 2019, a neuroscientist from the *Center for Neurotechnology at Columbia University (USA),* Rafael Yuste, pioneered a technology that can read and write to the brain with unprecedented precision without the need for surgery, but requiring genetic engineering. Following genetic manipulation, mice neurons are enabled to both become sensitive to and emit light, and, as a result, the researchers can both decipher what's happening in the animal's brain and write to it with an accuracy impossible with other techniques, thereby controlling the mice' neural activity. According to linguists, it will be just a matter of time before we can treat human patients with neurorehabilitation the same way we have been able to treat these animals.

2.2.4 The emergence of brain engineering and neurorights

The application of curated data and the ability to nudge people has driven the emergence of what is known as neuromarketing, influencing people to buy certain products and services. Likewise, it is possible to apply a neuromarketing algorithmic approach to other more sensitive topics, such as politics, which may lead to the violation of fundamental human rights: Neurorights are not a thing of science fiction anymore (Iberdrola, n.d.). Determining the ethical limits of neurotechnology could pose a threat to basic human rights, thus engendering the concept of neurorights. The increasing need for neurorights can be seen in five thematic areas as follows:

- Personal identity,
- Free will,
- Mental health,
- Equal access, and
- Protection against exclusions.

Some human rights could be diluted and weakened by the development of brain engineering. The Spanish government proposed new rules in November 2020 for regulating emotional AI that include specific provisions for neuro-rights. This is the first-time humans have ever had access to the contents of a person's mind, said Enrique Yuste (Asher-Schapiro, 2020). According to Strupp (2022), a recent ruling by the European Court requires now companies to protect data that indirectly relates to sensitive information, such as sexual orientation or health, which means that companies will face increasing legislative and regulatory pressures to apply special protections to data that were previously not considered sensitive—and biometrically inferred data will likely be seen as highly sensitive going forward.

2.3 Contemporary neuroethics principles

In addition to the AI ethics principles baseline detailed in the previous section, we studied contemporary ethics research in the neuroscience domain to better understand which additional ethical considerations are required within the context of socially aligned networks and the increasing use of emotional AI.

2.3.1 Emergence and development of neuroethics

According to Hrincu et al. (2021), while the field of neuroethics formally emerged on the back of an international conference in 2002, studies of ethical, legal, and societal impact posed by neuroscientific research focused on understanding the human nervous system, consciousness, and cognitive features of individuals, including thoughts, personalities, preferences, emotions, and behaviors have been undertaken for centuries. Hrincu et al. (2021) further introduced "Four pillars of Neuroethics" with ethical elements for consideration that are important from a socially aligned network perspective: (i) the self, (ii) social policy, (iii) ethics and neuroscientific practice, and (iv) public discourse.

While the latter two pillars are of particular interest from a neuroscience perspective, the former two raise important neuroethics concerns that we consider relevant within the context of socially aligned networks.

- **The self**—an individual person's freedom to perform actions leading to particular effects—may be constrained by socially aligned networks and emotional AI ecosystems to undermine control over action. The paper states that brain studies on "the biological correlates of mental and behavioral states reveal social judgment, emotion, consciousness, and other complex, multidimensional processes relevant to agential action." Further, the study highlights processes beyond conscious awareness that can affect how people act and orient themselves in the world, and those processes are suspected to impact personality, and the self.

- **Social policy**—concerns societal, legal, and policy implications of innovations with socially aligned networks and use of emotional AI. Relevant thematic sub-sets include personal and neuro privacy, legal consequences of behavior, access to innovation, and disparities caused by limited access to technology such as VR/AR/XR glasses. Similar to the neuroethics domain, our study expects an intermingling of socially aligned networks and the legal system, particularly in terms of behavioral research including (i) the prediction of future user behavior or users' mental abilities, (ii) detection of user's current mental states, and (iii) monitoring of human thought and consciousness states.

2.3.2 A neuroscientific ethics toolbox

Although there is continuing dissent with neuroscience researchers as to what constitutes an ethical use of some of those technologies, there have been attempts to build an "ethics toolbox" (Farah, 2015) aimed at recognizing, analyzing, and communicating ethical issues that arise within the neuroscience domain and we felt compelled studying those further.

Furthermore, (Farah, 2015) suggests that there are two fundamental ethical compartments when trying to identify useful ethical tools: consequentialist and deontological approaches.

- **Consequentialism** is introduced as an ethical framework where an act can be judged right or wrong depending on the expected values of its outcomes. It is most closely associated with philosophers Jeremy Bentham and John Stuart Mill.
- **Deontology** stems from the Greek word for "duty," and this philosophical approach determines ethicality in relation to a set of moral principles that specify personal obligations, duties, and rights. It is often associated with the 18th century philosopher Immanuel Kant.

Neither of those approaches are perfect and philosophers have often struggled when attempting to reconcile the two approaches, for example "by considering the beneficial consequences of recognizing rights" (Farah, 2015).

2.3.3 Philosophical roots for neuroethical decision-making

Within both philosophical roots, (Farah, 2015) introduces the following principles for ethical decision-making within a neuroscience context.

2.3.3.1 Deontology compartment

- **Personhood**—relates to the cognitive wherewithal (or potential for the not yet mature person) to think and act morally, with other philosophers also expanding this to rationality and self-consciousness. From a socially aligned network perspective, emotional AI ecosystems may need to consider persons' thoughts and the freedom of thought as personhood-related protected data category (see also the next section with case studies).

- **Dignity**—this principle was introduced into ethics with Kant's aim to distinguish and differentiate persons from objects, whereas objects have prices and are usually replaceable by paying a fair market price, persons are not replaceable. Kant called this for persons a "worth beyond value" which he subsequently termed dignity. Within the socially aligned network context, it could be argued that humans should not be objectified, and their unique data points as well as biometrically inferred data should not be treated as tradable assets within an emotional AI ecosystem context.
- **Commodification**—this principle follows on from the previous one and refers to the extension of market value to individual persons and their mental and cognitive capabilities. In other words, socially aligned networks and emotional AI ecosystems that commodify both personal data and data inferred from personal data can be seen as disrespecting human dignity. Further, it could be argued that an ethical use of such data within a socially aligned network context should lead to a fair compensation of the person for use of their data. Economically, this may not be feasible for commercial emotional AI ecosystems; however, it may still be the right thing to do ethically and morally.
- **Rights**—ethical principles referring to "rights" are moral entitlements and according to the US declaration of independence are "inalienable" from persons. One such example is the right to privacy as well freedom of speech (and arguably, freedom of thought and, more recently, brain rights).

2.3.3.2 Consequentialism department

- **Interests**—which can be seen as the counterpart to rights but missing the obligatory nature of those. And interests can often be weighted relative to each other.
- **Externalities**—are a term coined by economists referring to the effects of actions by one party on others who are not directly involved. Often some of those effects may be unintended yet still have a real impact on other parties. As shown in the case study section, socially aligned networks and ecosystems that rely on emotional AI, the use of algorithms

can work as a force-multiplier with both hugely beneficial and severely detrimental effects, which need careful and continuous consideration.

- **Sentience**—this principle within a consequentialist context, assumes that an entity must be sentient, that is capable of experiencing perceptual and affective states, which consequently may also include all animals.

2.3.3.3 Combined approaches

- **Subject autonomy**—when mixing both consequentialist and deontological considerations within a socially aligned network context, the treatment of human subjects will need to be guided by the risk-benefit ratio for deploying an emotional AI system as well as informed consent provided by the subject, respecting the subjects' autonomy in providing such consent.
- **Nonmalfeasance**—is a particularly important principle for subjects that may lack the competence required to provide informed consent, whereby regulations then often focus on protecting the person from harm by the socially aligned network and/or an Emotional AI ecosystem.

2.3.4 Neuroethics guiding principles

Our research has also looked closely to guiding principles in the realm of neuroethics, an important area in the field of neuroscience research focused specifically on ethical, legal, and societal implications.

Greely et al.'s (2018) Neuroethics Guiding principles published in *The Journal of Neuroscience* provide a set of ethical principles, which, when transferred and applied to socially aligned networks and emotional AI suggest the following.

- **Safety assessments are paramount**—as highlighted in the case studies section below, users' thoughts, consciousness, and neural data are exposed to algorithmic analysis within socially aligned networks. Consequently, users of such networks should place the highest priority on their own safety when using these emotional AI ecosystems including physical, psychological, and emotional consequences in the short, intermediate, and long term. Whereas physical safety will become increasingly important with the proliferation of connected devices, the Internet-of-Things (IoT) and tools such as VR/AR/XR-goggles, the emotional safety and mental well-being applies regardless of the devices that are used to access and interact with those ecosystems.
- **Anticipate special issues relating to users' capacity, autonomy, and agency**—responsible socially aligned networks must cater not

only for competent and autonomous adults but equally provide for people with diminished or developing autonomy and decision-making capacity. Although the ethical requirements for these emotional AI ecosystems that allow participation of those with limited, different or fluctuating capacity to consent are not new, they will require constant care and attention from the socially aligned network provider. This includes special consideration for vulnerable users, similar to banks and financial institutions being required to identify and assist in case of vulnerabilities of any kind. Naturally, this means vulnerabilities identified by the ecosystems' emotional AI must not be exploited for the profit of the socially aligned network provider.

- **Privacy and confidentiality protection of users' neural and/or thought-related data**—Users of socially aligned networks reasonably expect that the socially aligned network will protect both the privacy and confidentiality concerning their neural and thought-related data and its interpretation and inferences by emotional AI, which might include perceptions, emotions, memories, thoughts, and consciousness. Any such data should be treated as private, sensitive information and its collection, transmission, and storage should adhere to best practices for security and encryption. Where conflicts between privacy/confidentiality and data sharing arise, the former should always take precedence and be protected accordingly.
- **Predict and prevent plausible misuse of neural/thought-related user data**—Innovative and novel tools such as emotional AI and machine-learning algorithms applied to neural and thought-related data can be used for both good and bad means. Socially aligned network providers should be extremely mindful of possible misuses that might range from intrusive surveillance of thoughts, brain, consciousness, and mental states to incapacitate or impermissibly alter a person's behavior. Ecosystem providers have a responsibility to try to predict plausible misuse and ensure that such foreseeable risks are thoroughly understood and mitigated for.
- **Prevent premature tool adoption**—where socially aligned networks deploy innovative emotional AI algorithms the likelihood of individuals benefitting from those may be low and uncertain, whereas risks may be significant. Consequently, tech companies deploying algorithms that use neural and thought-related data should identify and mitigate related risks throughout the development and deployment lifecycle. This includes avoiding premature wide-spread adoption of such algorithms before the

risks are known and weaknesses of these tools have been fully understood.

- **Identify and address specific concerns of the public about neural/thought-related data**—although individuals' sensitivities over use of neural and thought-related data will likely vary between different cultures, geographies, and genders, public fears concerning the use of such data—whether justified or not—can have an important impact on the perception of big tech and ecosystem providers. For instance, a fear of mental invasion reaches far back historically and similarly does the notion of freedom of thought. Big tech firms, socially aligned networks and ecosystem providers should try to identify and address issues arising from the use of emotional AI that the public may find sensitive and off-limits.
- **Encourage public education and dialogue**—trust both by the public and individual users of socially aligned networks is an extremely precious commodity. One way for tech companies to build and retain that trust is by keeping both public and users transparently informed about how their neural and thought-related data are being used. Transparency is what ForHumanity and Carrier (2021a) suggests as basis to construct an "infrastructure of trust", which is crucial. And this should be easily achievable through a wide array of ways and means to communicate including public talks, user consultations, online scholarship, participating in interviews with Q&A sessions, and so on. Most importantly maybe, it requires a culture of openness and willingness to enter an active dialogue with the ecosystem's user base as well as the introduction of frequent, independent audits of such emotional AI systems.

2.4 Recent developments

2.4.1 Brain rights and legislative considerations

There are numerous examples in several jurisdictions where governments have been rolling out brain rights. Chilean legislators passed a new bill in September 2021 establishing the legal rights to individual identity and free will as well as mental privacy, becoming the first country to legislate neurotechnology that can influence a person's mind (The Neurorights Foundation, 2021).

Post "inception," professionals in Chile changed the debate regarding safety of burglar alarms to safeguarding the most valuable real estate on the planet—human minds. Chilean officials' goal is to empower the "cognitive autonomy" of its citizens by adjusting a constitutional reform that forbids technology from harming people's minds without their consent.

Experts are particularly concerned about the ability for periprosthetic implants to adversely affect freewill and liberty. The amendment to Chile's Constitution intends to provide the mind for the first time in history with nonmanipulable rights to guard it against advances in neurosciences, AI and emotional AI. The act protects the right to mental privacy, personal identity, freedom of thought, and equitable use of technologies that promote a sense of belonging (Seshadri, 2021). Legislation about neuroscience calls for clear boundaries and regular rules for protecting and maintaining the mind and physical body. Cultural respect for two separate privacy policies involving information and self-governance is important to legislate for the individuals. Informational privacy, or the protection of a person's physical or psychological identity with respect to other individuals, is particularly critical.

The current rights regarding informational privacy are intended to promote individual autonomy from the unfair dissemination of personal material, in addition to helping to protect the freedom of one's mind. For example, in 2018, the Court of Appeal for England and Wales found a right to privacy, or the psychological integrity of a person, protected. This intersection between an individual's privacy and integrity allows them to decide the kind of right they want to claim as and when required. Informational self-determination then refers to the right for individuals to determine what information is visible to others beyond their sphere of influence. Ostracized from the concept of informational self-determination is the liberty that each individual should have self-determination. With present-day technology, neuroscientists cannot conclusively distinguish an individual's intentions from the electrical activity recorded in their brain. Thus, brain signals are now being processed with the aid of AI, more specifically emotional AI.

2.4.2 Algorithmic limitations and cognitive freedom

Self-learning technology utilizes algorithms which cannot predictably be determined and are often limited by too coarse definitions. This leads to an onus on a human being, to bare their innermost consciousness, as the machine performs his or her job via an algorithm. In addition to improving our brain functions, entrepreneurs like Elon Musk have devised new technological innovations such as Neuralink aimed at enhancing human cognition (Neuralink.com, n.d.).

Our brains are being increasingly enhanced by neurotechnology, and these tools are becoming more widespread. Liberty to make choices with regard to one's own cognition is considered to be an idea of what is called cognitive liberty. Indian courts have repeatedly recognized freedom of thought as a constitutional guarantee. Such freedom of thought is likewise acknowledged to always embody the concept of finding cognitive freedom (Garg, 2022).

The aim of creating common cognitive discourse in the brain—AI interface is not to obtain agreement among the community regarding moral rightness. To the contrary, disagreement may boost your reasoning abilities. The ideal response of the brain—AI ethics community to developments in neuroscience and AI is not a general consensus, but instead knowledge sharing through dialogue and collaboration (Ienca, 2019).

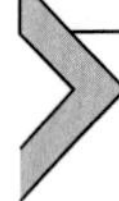

3. Human—AI interfaces within socially aligned networks

3.1 Case studies

Ethical studies of human—AI interfaces within a socially aligned network context require investigating case studies of instances where users were subjected to the deployment of emotional AI algorithms and to consider whether neuroethics alone would be sufficient to address these ethical challenges or if there were any gaps within contemporary thinking and approaches. The following section explores a number of case studies and highlights areas within the emotional AI spaces that will require careful risk assessment and cost-/benefit-analyses when deploying those methods. Further, it is also worthwhile keeping in mind that just because something can technically be done (and easily/cheaply), it may not be ethical and could be morally wrong.

3.1.1 Increasing extraction, derivation, and amplification of human thoughts

Alegre (2022) postulates in "Freedom to Think" that "*Big tech is reading us all everywhere, all the time. And the potential to scale up attacks on our freedom of thought is almost limitless. Once we lose our freedom of thought, we lose what it means to be human.*" Dystopian as it may sound, Alegre's view can easily be extended to socially aligned networks and the way tech companies behind those ecosystems deploy AI to commercially exploit users via the human—AI interface.

While users of socially aligned networks are not deliberately and consciously sharing their thoughts with these Big Tech companies, the thoughts are being extracted, derived, and amplified by deploying emotional AI systems aimed at deriving metadata based on both their online activity as well as inactivity and content that users delete—all for commercial gain of the tech companies behind these ecosystems.

3.1.2 Big Tech's (ab)use of human self-censorship

This is not a particularly new trend as evidenced in the 2013 paper "Self-censorship on Facebook" (Das & Kramer, 2013) featuring a study of almost four million Facebook users concluding that any content that users are entering into a comment box and then either subsequently delete or alter prior to publication is not private as it is stored within Facebook's systems and naturally also available for analysis by Facebook's AI that can make industrial scale automated inferences about every users thoughts and opinions.

3.1.3 Nonconsensual personality profiling

Further, when combined with smartphones and connected devices, user interactions can easily be used to collect extensive data sets that then serve for behavioral analysis with the help of emotional AI systems that pose serious threats to individuals' privacy as a study by Stachl et al. (2020) finds: their analysis of a large sample examined the extent to which individuals personality traits including communication and social behavior, music consumption, app usage, mobility, overall smartphone activity, and day-/night-time activity and concluded that users' personality can be predicted with a fairly high level of accuracy.

Whereas humans can normally decide which part of their self they want to reveal about their own thoughts, feelings, and consciousness, socially aligned networks that are making these interferences without consent, let alone informed consent—as often users are not aware what inferences have been made by the ecosystems at all—represents in effect a violation of human rights including the rights to freedom of thought and expression.

3.1.4 Mass manipulation and alteration of emotional states

Another 2-week Facebook experiment from 2014 targeting 700,000 Facebook users and resulting in a study entitled "Experimental evidence of massive-scale emotional contagion through social networks" highlighted that users' emotional states could be altered by manipulation of their news-feeds (Kramer et al., 2014).

3.1.5 Deliberate targeting and abuse of vulnerabilities

To add to both previous case studies, reading users' minds and influencing their thoughts and behaviors, according to another leaked Facebook document analyzed by The Australian newspaper, the social media company allegedly also offered real-time insights into users' emotional states with the aim to making this information available for targeted advertising based on the use of AI and algorithmic recognition of users' emotional states. The internal report stated that Facebook can monitor posts in real-time to determine when teenagers feel "stressed," "defeated," "overwhelmed," "anxious," "nervous," "stupid," "silly," "useless," and a "failure" (Levin, 2017)—in short, when they are most vulnerable.

Big tech companies running social media and search platforms such as Facebook, Google, Instagram, TikTok, and Twitter have an economic interest into online advertising. Users' interactions with technology and the information that Big Tech companies derive from those—increasingly with the use of emotional AI and algorithms—is driving real-time-bidding (RTB) by making it easy to purchase all of this information. As Reset Australia highlights in a 2021 study (Williams et al., 2021), social media platforms profile young people for advertising purposes and the study's conveners have been able to acquire targeted advertising slots based on identified vulnerabilities ranging from extreme weight loss, gambling, and online dating with older wealthy men in teenagers aged between 13 and 17.

3.1.6 Prediction of political orientation threatening civil liberties

The increasing use of AI and algorithmic profiling goes far beyond selling of advertising space. As (Kosinski, 2021) notes in a 2021 study, with the use of facial image recognition technology and algorithms, it is possible to predict the political orientation of a social media user purely based on a photo of their face with a high accuracy even when controlling for age, gender, and ethnicity. Kosinski (2021) concludes that his "findings have critical implications for the protection of privacy and civil liberties."

3.2 Recommendations

We have recently seen an increasing trend of large and substantial regulatory fines issued to Big Tech companies, often for deploying AI in a way that may have intervened or breached individual data privacy and/or data protection legislation. The damage of such breaches goes far beyond purely financial penalties; it also has an impact on the organization's reputation, standing with the relevant regulatory authorities, user base, and society.

Consequently, we recommend that Big Tech companies who deploy some or all of the methods highlighted in above case studies to analyze, review, document, and internally question their operational processes and use of emotional AI systems with an ethical lens.

In addition, we recommend firms who use emotional AI systems to establish ethics committees as explained by ForHumanity and Carrier (2021b) with diverse multistakeholder input (ForHumanity and Carrier, 2022) as well as deploying frequent (at least annual) independent audits of AI systems (ForHumanity and Carrier, 2021a).

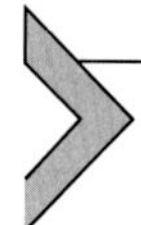

4. Boundary issues

4.1 Morality, ethics, and spirituality

Spiritual or moral enhancement calls for logical participation of a data-driven self; genetic enhancements will often fall short. Does it make sense that AI was created intentionally vis-a-vis spiritual or moral elevation (Peters, 2022)?

When we hear the word "spiritual," our brain automatically makes us think about religion. But why is this so? We are yet to understand the true effects of spirituality, which has everything to do with inner peace and maybe human understanding of attaining salvation. Is it important to make machines understand what spirituality is, what morality means, what ethics constitutes, or what religion is all about? Because every human being has their own individual understanding of the same issue.

Will spiritual AI be the solution to our problems that we humans have carefully developed throughout evolution? For most of us, even morality and ethics have a different meaning as one may say that morality comes from the heart and ethics from the brain. Is singularity going to arrive soon? No one really knows, but we can definitely explore and have an open discourse about it; no one is stopping us, definitely not the Law.

4.2 Immature legal frameworks

The absence of mature legal frameworks currently around neuroscience is going to play a big role in determining the after-effects of the criminal justice system. The laws protecting neuroscience additionally pose an issue regarding consent. The proposed actions need to be accepted, which is to say that consumers and users must have proper information about what amount of data are being used and the possible consequences and effects it may have. Thus, having access to this data is a prerequisite. It is imperative that everyone should have the opportunity to benefit from neurotechnologies, and there ought to be fair access to these materials including biometrically inferred data. But what is clear currently is that the legal frameworks on neurotechnologies are not readily available and sufficiently evolved to cater for these emerging challenges.

Should we consider our interconnected social and environmental network, which encompasses living beings and nature, as well as the possibility of robots and machines that possess human-like thinking and behaviour, and could potentially even qualify as sentient artificial intelligence in the near future? If so, we must also ponder upon the idea of having human emotions, feelings like sorrow, happiness, and sometimes blank thoughts in algorithmic forms as well. Evidence of such can be checked upon several investigations and research going on around the world and some thought-provoking individuals thinking of establishing new settlements and human colonies in other parts of the universe away from Earth.

As evident with the case studies in this chapter, the notion that Big-Tech companies in the social and socially aligned network realms are using AI to both determine mental states and thoughts as well as influencing their emotions and behavioral responses is not particularly new.

4.3 Ethical boundaries for IoT devices

With the increasing merging of Internet-of-things (IoT) devices and growing integration of connected smart devices into our daily lives, it is important to identify and set ethical boundaries and challenge Big-Tech companies on the otherwise careless deployment of smart devices and abuse of humans' neural and thought-related data. As Blackman (2022) suggests in *Ethical machines*, the questions go far beyond pure definitions of what "ethics" is and should be focusing on "what ethics is about." In the study for this chapter, we aimed to look beyond the consensus ethical baseline within an AI setting and instead aimed to understand scientific purist approaches where thoughts and neural data are concerned.

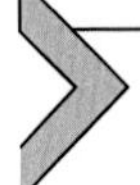

5. Conclusion

5.1 Neuroethics—a suitable blueprint

In our view, the neuroscience domain, and within it the Neuroethics branch, represents best practice in a domain filled with substantive ethical challenges. It may not be perfectly transferable, but we think that given the importance surrounding human consciousness, freedom of thought and related human rights, Neuroethics approaches provide a solid foundation and safeguarding features.

Consequently, both socially aligned networks including some of the recent emerging ecosystems for crypto-/NFT-technologies as well as metaverses and augmented reality realms as well as algorithmic processing within the emerging field of quantum-computing means that this will remain a highly dynamic and fast-paced environment—requiring philosophy and ethics to keeping up the pace.

5.2 Regulation alone is likely not enough

Although regulation (see Chile's neurorights) is gradually catching up, we remain doubtful that regulation alone will ever be able to level the playing field for socially aligned networks. This can then only be counterbalanced by continuous ethical research as well as an inclusive discussion of those big ethical questions with the general public and all stakeholders concerned, often referred to as "Diverse inputs and multi-stakeholder feedback" (ForHumanity and Carrier, 2022).

We hope that this chapter will help open up new perspectives, ultimately leading to a better understanding of ethical challenges within the context of socially aligned networks. Equally, we hope that our study may inspire further multidisciplinary research in this relatively young branch of ethics and behavioral philosophy.

5.3 Generative AI

Finally, with increasing proliferation of GPT-x and other Large Language Model (LLM)-based chatbots such as ChatGPT (Open AI), Bard (Google), Ernie (Baidu) and similar tools currently under development by Amazon, Meta, Apple, Alibaba, JD.com, and WeChat/Tencent, it is only a matter of time when human-to-human interactions may also be frequently replaced by conversations with those bots.

User responses to those conversational AI tools have been greatly polarized painting both dystopian and utopian scenarios. And this will likely continue with a wave of new generative AI tools that are capable to turning user prompts into pictures as well as hyper-realistic videos.

Given how far and deep AI has already penetrated society and how quickly LLMs are evolving, it will become increasingly important for humans to understand who (e.g., another human or chatbot) they are corresponding with and develop a practical sense of both capability and limitations of such AI.

From both a social and socially aligned network perspective, there is a chance that generative AIs will take over and will largely replace human network participants with social chatbots, in which case we feel it could be beneficial to identify in all correspondence whether the opposite is an actual human or an AI tool. This is to ensure transparency in any communications, whether they are Human-to-human or human-to-machine and also provide ultimately the individual person as well as humanity collectively with the ultimate control to discontinue any human-to-machine interaction should they choose to do so (e.g., human-in-the-loop).

Society is at a cross-road and facing grave danger if trust—one of humanity's oldest principle of peaceful living together—is undermined by an emergence of deepfakes and digital twinning by AI. If anything, an individual natural person's unique identity must be protected at all costs and with similar, if not more stringent measures that we see in the financial markets countering identity theft. For instance, this does not mean that users should not be able generate their own avatars in the metaverse or applications such as spatial, but if they do, they should be clearly identified as a such—more so if they supercharge those with their own persona by using generative AI.

5.4 Future research

We suggest the following areas for future research and investigation where emotional AI and human–AI within the context of socially aligned networks are concerned.

- Investigate the proportion and trends affecting digitally excluded parts of the world.
- Investigate reasons of why segments of the global digital population are not on social media networks.
- Investigate sentience of machines, robots, and AI.
- Investigate whether neuroethics aimed at noninvasive brain stimulation differs (or not) to invasive brain stimulation and if/how any of these ethical approaches may similarily apply to emotional AI.
- Investigate how independent audit approaches in financial markets can be applied to independent audit of emotional AI systems.
- Investigate Big Tech firms' internal governance structures and usage of diverse multistakeholder inputs, ethics and risk committees and these entities' medium- to long-term performance in comparison to others lacking those governance instruments.
- Investigate the interplay of generative AI within the context of socially aligned networks.

- Investigate the impact of generative AI on human thinking and neurological implications when interaction with AI.

Acknowledgments

We are grateful to Dr. rer. nat. Moritz Julian Maier, Fraunhofer Center for Responsible Research and Innovation (CeRRI), Berlin for sharing his insights and observations as part of a multiyear research program with us. You can find out more about this project here: https://www.cerri.iao.fraunhofer.de/en/projects/stimcode.html.

References

Ahluwalia, M. (2021). Legal governance of brain data derived from artificial intelligence. *Voices in Bioethics, 7*. https://doi.org/10.52214/vib.v7i.8403

Alegre, S. (2022). *Freedom to think*. London, England: Atlantic Books.

Asher-Schapiro, A. (2020). Thomson Reuters Foundation. *This is not science fiction," say scientists pushing for "neuro-rights"*. https://www.reuters.com/article/us-global-tech-rights-idUSKBN28D3HK. (Accessed 17 August 2022).

Blackman, R. (2022). *Ethical machines*. Boston, MA: Harvard Business Review Press.

Custers, B., & Fosch-Villaronga, E. (2022). Law and artificial intelligence. In *Information technology and law series* (1st ed.). The Hague, Netherlands: T.M.C. Asser Press.

Dartmouth Undergraduate Journal of Science. (2013). *Neurocriminology: The disease behind the crime*. Dartmouth Undergraduate Journal of Science. https://sites.dartmouth.edu/dujs/2013/11/19/neurocriminology-the-disease-behind-the-crime/. (Accessed 17 August 2022).

Das, S., & Kramer, A. (2013). Self-censorship on Facebook. In *Proceedings of the 7th international conference on weblogs and social media*. ICWSM.

Dixon, S. (2023a). Number of social media users worldwide from 2017 to 2027 (in billions) [Graph]. In *Statista*. https://www-statista-com.ezproxy-s2.stir.ac.uk/statistics/278414/number-of-worldwide-social-network-users/. (Accessed 17 July 2023).

Dixon, S. (2023b). Social network penetration worldwide from 2018 to 2027 [Graph]. In *Statista*. https://www-statista-com.ezproxy-s2.stir.ac.uk/statistics/260811/social-netwo rkpenetration-worldwide/. (Accessed 17 July 2023).

Ethics Washing Is When Ethics is A Substitute for Regulation. (2018). *Data ethics*.

Farah, M. J. (2015). An ethics toolbox for neurotechnology. *Neuron, 86*, 34–37. https://doi.org/10.1016/J.NEURON.2015.03.038

Fjeld, J., Achten, N., Hilligoss, H., Nagy, A., & Srikumar, M. (2020). Principled artificial intelligence: Mapping consensus in ethical and rights-based approaches to principles for AI. *SSRN Electronic Journal*. https://doi.org/10.2139/ssrn.3518482. https://doi.org/10.2139/ssrn.3518482, 2020. (Accessed 17 August 2022)

ForHumanity, & Carrier, R. (2021a). *Infrastructure of trust for AI — guide to entity roles and responsibilities*.

ForHumanity, & Carrier, R. (2021b). *The rise of the ethics committee*.

ForHumanity, & Carrier, R. (2022). *Diverse inputs and multi stakeholder feedback (DI and MSF)*. ForHumanity. https://forhumanity.center/article/diverse-inputs-and-multi-stakeholder-feedback-di-msf/, 2021. (Accessed 17 August 2022).

Garg, I. (2022). The Vidhi Centre for Legal Policy. *The time is now for a 'neuro-rights' law in India*. https://vidhilegalpolicy.in/blog/the-time-is-now-for-a-neuro-rights-law-in-india/?utm_source=rss&utm_medium=rss&utm_campaign=the-time-is-now-for-a-neu ro-rights-law-in-india. (Accessed 17 August 2022).

Greely, H. T., Grady, C., Ramos, K. M., Chiong, W., Eberwine, J., Farahany, N. A., Johnson, L. S. M., Hyman, B. T., Hyman, S. E., Rommelfanger, K. S., & Serrano, E. E. (2018). Neuroethics guiding principles for the NIH BRAIN initiative. *Journal of Neuroscience, 38*(50), 10586. https://doi.org/10.1523/JNEUROSCI.2077-18.2018

Hrincu, V. M., Courchesne, C., Lauc, C., & Illes, J. (2021). Contemporary neuroethics. In *Encyclopedia of behavioral neuroscience* (2nd ed.). https://doi.org/10.1016/B978-0-12-809324-5.24095-9

Iberdrola S.A., n.d. What are neurorights and why are they vital in the face of advances in neuroscience. https://www.iberdrola.com/innovation/neurorights#:~:text=Neurorights%20can%20be%20defined%20as,its%20activity%20as%20neurotechnology%20advances (Accessed August 17, 2022).

Ienca, M. (2019). *Neuroethics meets artificial intelligence*. The Neuroethics Blog. http://www.theneuroethicsblog.com/2019/10/neuroethics-meets-artificial.html. (Accessed 17 August 2022).

Kerrigan, Charles (2022). *Artificial intelligence: Law and regulation*. Edward Elgar Publishing. https://doi.org/10.4337/9781800371729

Kosinski, M. (2021). Facial recognition technology can expose political orientation from naturalistic facial images. *Scientific Reports, 11*. https://doi.org/10.1038/s41598-020-79310-1

Kramer, A. D. I., Guillory, J. E., & Hancock, J. T. (2014). Experimental evidence of massive-scale emotional contagion through social networks. *Proceedings of the National Academy of Sciences of the USA, 111*. https://doi.org/10.1073/pnas.1320040111

Levin, S. (2017). *Facebook told advertisers it can identify teens feeling "insecure" and "worthless*. The Guardian.

Lin, P. K. (2021). *Machine see, machine do : How technology mirrors bias in our criminal justice system*.

Neuralinkcom, n.d. NeuraLink. https://neuralink.com/ (Accessed August 17, 2022).

OECD - Recommendation of the Council on Artificial Intelligence. (2019).

Peters, T. (2022). Will superintelligence lead to spiritual enhancement? *Religions, 13*. https://doi.org/10.3390/rel13050399

Routley, N. (2022). *Visualizing the world's top social media and messaging apps*.

Seshadri, N. (2021). *Chile becomes first country to pass neuro-rights law*. https://www.jurist.org/news/2021/10/chile-becomes-first-country-to-pass-neuro-rights-law/. (Accessed 17 August 2022).

Stachl, C., Au, Q., Schoedel, R., Gosling, S. D., Harari, G. M., Buschek, D., Völkel, S. T., Schuwerk, T., Oldemeier, M., Ullmann, T., Hussmann, H., Bischl, B., & Bühner, M. (2020). Predicting personality from patterns of behavior collected with smartphones. *Proceedings of the National Academy of Sciences of the USA, 117*. https://doi.org/10.1073/pnas.1920484117

Strupp, C. (2022). EU court expands definition of sensitive data, prompting legal concerns for companies. *The Wall Street Journal*.

The Neurorights Foundation. (2021). *NeuroRights in Chile*. https://plum-conch-dwsc.squarespace.com/chile. (Accessed 17 August 2022).

United Nations Population Fund (UNFPA). (2022). *United Nations Population Fund (UNFPA), UNFPA Population data portal*. United Nations Population Fund (UNFPA). https://pdp.unfpa.org/. (Accessed 17 August 2022).

Williams, D., McIntosh, A., & Farthing, R. (2021). *Profiling children for advertising: Facebook's monetisation of young people's personal data*.

XR Safety Initiative, n.d. Biometrically-inferred data (BID) https://xrsi.org/definition/biometrically-inferred-data-bid (Accessed August 17, 2022).

CHAPTER SEVEN

Emotion AI in healthcare: Application, challenges, and future directions

Kriti Ahuja
Department of Allied Sciences, School of Health Science and Technology, University of Petroleum & Energy Studies, Dehradun, Uttrarakhand, India

1. Introduction

Artificial intelligence (AI) and related technologies are coming in use exponentially. They have been proved to be of great help in all the fields especially healthcare. One of the founders of AI and modern computers was Alan Turing in the year 1950. He is known for his "Turing Test" that revolves around the fact "When can a machine be called intelligent?" Alan Turing devised a test in which a man and a woman are put in different rooms. An interrogator is put in such a way that he/she cannot see the man or the woman. All three elements are connected through teletyping, that is, a kind of an "electronic" chat system. The interrogator is free to ask a variety of questions in order to accurately guess the identity of the man and the woman. Woman answers the questions accurately while the man fakes his identity in order to confuse the interrogator. Later, the man is replaced by a machine. If the machine passes the test and is able to fool the interrogator that it is a woman, the machine is called "intelligent." The test is termed as the imitation test (French, 2000).

Turing's imitation game was the base of AI. It directed humankind to make algorithms for computing machines to be able to behave like humans in certain aspects making it almost impossible for the interrogator to differentiate between the machine and human (Fazi, 2019).

In modern times, the Turing test has been adapted in various forms. An example can be when a human or a machine is kept in a single room, and the interrogator has to tell if the room has a human or machine. Despite these variations, the "Turing test" remains the core basis of AI (French, 2000). AI can be categorized in two major forms as shown in Fig. 7.1 (Ranade et al., 2018)—

Emotional AI and Human-AI Interactions in Social Networking
ISBN: 978-0-443-19096-4
https://doi.org/10.1016/B978-0-443-19096-4.00011-0

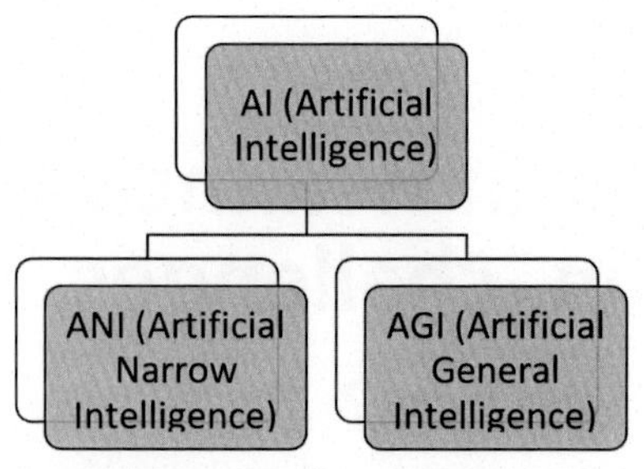

Figure 7.1 Categorization of artificial intelligence (AI) in terms of artificial narrow intelligence (ANI) and artificial general intelligence (AGI).

1. **Narrow artificial narrow intelligence (ANI)**: Making the machine intelligent comes under narrow AI. The resultant machine is capable of performing under the coded intelligence but within the domain it is set for.
2. **Broad artificial general intelligence (AGI)**: When the machine is capable of performing both inside and outside the domain, it comes under broad AI. A great amount of research is being done in this area. However, the present research is still far from these expectations.

2. Applications of emotion AI

Artificial intelligence is termed as emotion AI (EAI) when it can predict human emotions accurately using image, videos, audios, or texts as input (Ranade et al., 2018).

Apart from that, EAI remains to be an efficient and significant tool, which is changing the world due to its extremely important and wide applications in almost every sector (The Most Technologies That Uses Artificial Intelligence In USA, 2019/2019). The biggest dilemma in the concept is that usually we as humans do not show what we think and do not say what we feel. As a matter of fact, building a machine which overcomes this issue completely is a handful (Fölster et al., 2014). Some of these include business, retail, medical diagnosis, patient care, banking, advertisements, call centers, autonomous cars, car safety, job recruiting, and even fraud detections as represented in Fig. 7.2 (Ranade et al., 2018).

Some of these uses are discussed below in brief detail.

1. **Retail:** EAI has its significant application in the retail industries. Retails use EAI in their stores to detect the emotional state and moods of the customers coming in and leaving. The information is crucial and can help in growing the business.

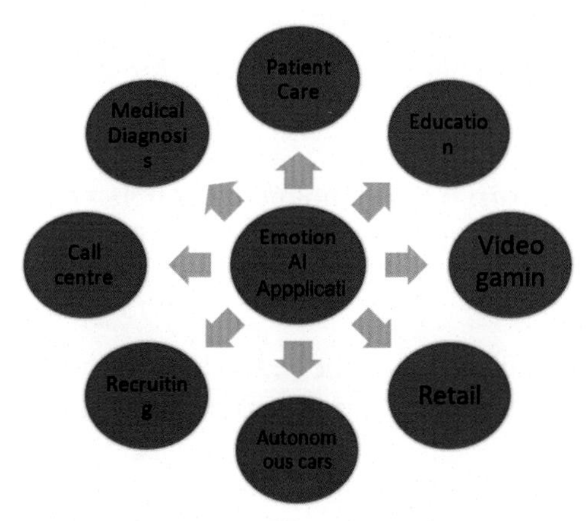

Figure 7.2 Applications of emotion artificial intelligence.

2. **Business:** Businessmen can not only maintain good terms with existing customers but also attract novel customers to their brands. EAI is capable of maintaining long-term connections with the customers. It allows this by making the business owner generous and considerate of customer's feelings toward the product under question (The Importance of Emotional AI in Business in the 21st Century, 2022/2022).
3. **Recruiting:** EAI can prove to be an excellent tool for recruiting liable and apt candidates for the job in question.
4. **Autonomous cars:** Car companies can potentially use EAI in their cars to predict the driving and riding experience of people while they take a test drive. This can increase the sales as well as help the companies make cars that are consumer friendly.
5. **Car safety:** Extreme emotions have been an alarming cause of road accidents in recent times. The EAI cameras and sensors in the car detect the extreme emotions or drowsiness in the driver and can hence alert the driver well in time.
6. **Call center routing**: A well-constructed EAI can be a boon to call centers as it can route the call to a well-trained employee or an intern detecting the mood of the customer.
7. **Education**: Learning software has been developed to detect kid's emotions. The software changes the tasks according to the child's emotions and reactions to the task at hand. The prototype observes the frustration or challenge in the face of the user to increase its efficiency.

8. **Employee safety**: Due to rising stress and anxiety in employees, employee safety has been the first concern for the industries. EAI analyzes the anxiety and distress levels in the employees especially those who have extreme jobs.
9. **Medical diagnosis**: A great amount of research is going on in the diagnostic fields. Doctors have been highly helped with software with voice and human image analysis in diagnosis of diseases especially highly risky mental disorders.
10. **Patient care**: Full time care for bedridden patients and patients in old age has been an important area since time immemorial. A "nurse-bot," an EAI not only monitors the patient's overall well-being but also reminds the patient to take their medications on time.

However, EAI is a fairly new concept and still under development. Presently, the areas in computer sciences that consider the concept of emotions in their research include the systems with emotion-based architecture and human—computer interaction. Hence, EAI is yet to tackle various challenges for its efficient use. One of these major challenges include integrating EAI in the healthcare sector.

3. Emotion AI and healthcare

As discussed in earlier, EAI is capable of reading human emotions such as anger, disgust, misery, happiness, and neutrality by measuring microexpressions that are tedious for human eye to capture (How Emotion AI Can Be A Game Changer In Healthcare | Enablex Insights, 2022/2022). In the last few years, mental health awareness has been a great agenda in the general society. This has led to increased services in its cure and also the removal of stigma that it surrounds. EAI is an emerging technology that has believed to have tremendous impact in healthcare especially in mental healthcare.

Latest study as referred to (Mental Health: Lessons Learned in 2020 for 2021 and Forward, 2022/2022) figured out that over one billion people in a seven billion global population suffer from mental disorders and in most of the low-income countries almost 75% of the sufferers remain untreated. So much so forth that every four-sixth of the minute you pass, someone commits suicide. The estimations are both alarming and of great concern. This makes it more important for a solution in this regard. As for depression, it is believed that it will be a major disease of concern in most countries in the upcoming 10 years (Mental Health: Lessons Learned in 2020 for 2021 and Forward, 2022/2022). EAI has known its applications in various fields including healthcare, especially mental healthcare. It offers doctors an efficient option to not only monitor the well-being of their existing patients but also diagnose doubtful cases accurately (Walsh et al., 2017a).

4. Applications of EAI in healthcare

EAI can be integrated in healthcare in various forms in order to benefit the society. Some of these forms are as visually represented in Fig. 7.3.

1. **Psychiatric counseling**: Studies have been done and a lot of devices under the name chatbots are being made in this domain. The devices are capable of counseling the patients almost as good as a human counselor and also observe changes in behavior, emotions, and daily life patterns of the patient in order to give more insight to the patient's health leading to his/her betterment (Oh et al., 2017).
2. **Pregnancy care**: Pregnancy has been known to have risk factors in terms of emotions and mental health such as anxiety, depression, or stress. An incredible enhancement and maintenance of health in pregnant women can be achieved through EAI. However, not much research was found in the identification of emotions in pregnant women. Hence, the research in this area is highly recommended (Oprescu et al., 2020).
3. **Diagnostic in mental health**: Mental health disorders can be diagnosed with much ease with the help of EAI. They have known to deliver the diagnosis way faster and way more accurately. EAI can help in provided better mental health facilities at lower costs (Walsh et al., 2017a).
4. **Treatment of Autism**: In total, 1.67% people in India suffer from autism, technically termed as autism spectrum disorder (ASD). People suffering from ASD have troubles in social, emotional, and communications skills so much so forth that it effects their lives. The added layer of EAI can help them communicate better with other individuals by

Figure 7.3 Applications of emotion artificial intelligence in healthcare.

helping them comprehend the emotions of the collocutor (How Emotion AI Can Be A Game Changer In Healthcare | Enablex Insights, 2022/2022).

5. **Nursing**: EAI has worked miracles in the field of nursing. Japan, being the largest producer of robotic nurses popularized as nursebots are considered of great benefit to the country. A nursebot can help in day-to-day tasks especially of the elderly and the specially abled and also engage them in small conversations (NurseBot - The Index Project, 2022/2022).
6. **Predicting suicide risks:** EAI can and does predict suicide risks by playing an important role by observing and hence analyzing the patient's behavior (Walsh et al., 2017a). As shown in the figure below, suicides have been a rather alarming cause of death, which makes its prediction way more important.

5. Case studies of EAI devices

Some major EAI products impacting the healthcare sector are discussed below.

5.1 Chatbot

Chatbot, now a popular prototype, is an EAI, which understands and sympathizes with your emotions by chatting with you. Chatbot is emerging to be used in developed countries in mental health departments for patients who need talk therapy. Chatbot is capable of interacting with the patient just as a human would. This makes it highly influential for patients who need therapies which are proved to be a time-intensive treatment. The most common functions of chatbot are to deliver therapy, train, and screen. Most of them take text as a source input and return outputs in the form of text, audio, or visual language. Most of the chatbot developed yet are focused mainly on autism and depression.

EAI chatbots depend on natural language processing (NLP) in order to figure out the question and its meaning and answer it correctly. The construction of any chatbot uses three main fundamentals, including natural language processing, AI, and machine learning with natural language processing being the groundwork of the invention. The four major concepts NLP, natural language understanding (NLU), natural language generation (NLG), and automatic speech recognition (ASR) guide the achievement of a chatbot (Abd-alrazaq et al., 2019). A clear representation can be seen through Fig. 7.4.

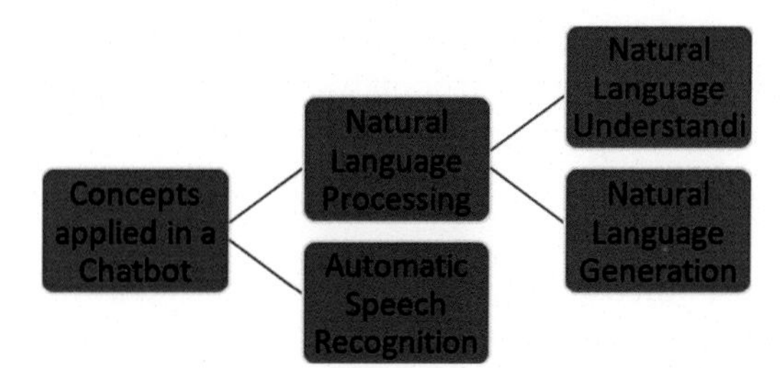

Figure 7.4 Major concepts used by a chatbot.

Natural language processing: Both Google and Siri use NLP as the foundation of speech recognition. It can be divided into two major stages—NLU and NLG. The five other steps that come under NLP, include lexical analysis, syntactic analysis, semantic analysis, discourse analysis, and pragmatic analysis (Ayanouz et al., 2020).

Natural language understanding: NLU is a comparatively quite gruelling. This is due to the complex structure of natural language. It converts formless data to a form that is understandable by the computer (Oh et al., 2017; Ayanouz et al., 2020).

Natural language generation: Later, NLG comes into play under which the text is linguistically realized, and an appropriate result is formed including the correct arrangement of words and phrases (Ayanouz et al., 2020).

Automatic speech recognition: ASR, also termed as computer speech recognition (CSR) or speech to text (STT), is a concept by which computers convert the user's speech into text. It is preceded by speaker identification. Speaker identification often eases the process of speech conversion. Hence, the systems which are trained are known as speaker-dependent systems, while others are known as speaker-independent systems (Ayanouz et al., 2020).

5.1.1 Advantages of today's chatbot (Abd-alrazaq et al., 2019)

Some of the advantages of the chatbot are discussed below.

1. Many patients would be more comfortable and will talk more freely to a machine rather than a human therapist.
2. It reduces manpower.
3. It is time-independent and can be used as and when the patient needs it.
4. It also adds wider accessibility to the picture.

5.1.2 Limitations of today's chatbot (Abd-alrazaq et al., 2019)

Some of the limitations of the chatbots are discussed below.

1. Chatbots fail to recognize grammatical errors.
2. The existing chatbots use an unfathomable interface which is not user-friendly.
3. The techniques used in building chatbots are fixed rule templates and straightforward machine learning.
4. Each language tends to have its own rules of usage of words and punctuations. Current chatbots fail to determine these differences.
5. Most of the chatbots today only accept inputs in the English language. This adds language barriers in terms of usage for non-English speakers.
6. A satisfactory level of accuracy is yet to be achieved in today's chatbots as most of them give an unpredictable response or out of context answers when the user changes the subject unexpectedly.

5.2 Nursebot

Nursebot is a fairly new concept popularized as the nurse robots or the robotic nurses. Nursebots are already being used by the hospitals in Japan and some other countries of the United States. The robotic nurses are being employed to help the elderly and the specially abled in the hospitals. Allocating nurse bots in the areas of use can help the companies to utilize their nurses in better and more important areas of work like the emergency rooms or rooms for critical care (Qaraqe et al., n.d.). Nursebots are a miracle as they are capable of helping the elderly in routine activities such as bathing, dressing up, reminding the medications, and helping them eat. Apart from that, they can often engage the patient in social activities by having small conversations with them (Machine Learning for Nursing – 8 Current Applications | Emerj Artificial Intelligence Research, 2022/2022).

Two incredibly popular robotic nurses are shown in Fig. 7.5 below.

1. **Personal Robotic Assistant for the Elderly (PEARL)**
   PEARL was developed in the United States by the University of Pittsburgh, Carnegie Mellon University, and University of Michigan.

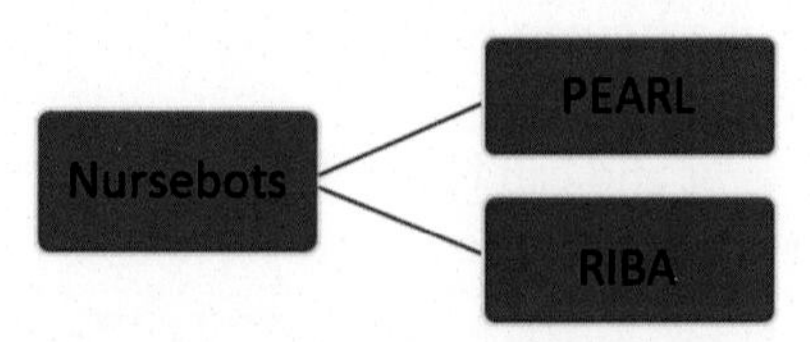

Figure 7.5 Two popular robotic nurses.

PEARL can be expanded as Personal Robotic Assistant for the Elderly (NurseBot - The Index Project, 2022/2022). The role of PEARL is not only to remind the patients about their routine activities but also to guide the older patients through their surroundings. It is an outstanding invention for the elderly and specially abled people to help them perform their day-to-day tasks both cognitively and physically. It has a sophisticated algorithm which also keeps tabs on people, the patient's habits, and how likely the reminder given is likely to be performed (Robotic Nurses | Computers and Robots: Decision-Makers in an Automated World, 2022/2022).

2. **Robot for Interactive Body Assistance (RIBA)**
   RIBA is a robotic nurse developed in Japan, the leading country in terms of production of nursebots to address various tasks in a hospital. RIBA stands for "Robot for Interactive Body Assistance." RIBA is developed in order to lift the patients and move them in and around rooms. It is capable of lifting up to 135 lbs. RIBA is known for its strong arms accompanied with high profile sensors in order to prevent slipping, tow cameras, and two microphones in order to follow the instruction given by its operator(Robotic Nurses | Computers and Robots: Decision-Makers in an Automated World, 2022/2022).

5.3 Coronavirus panic index

The COVID-19 pandemic had a deep impact on the society with significant loss of lives, loss of jobs, social, and financial restrictions leading to a great amount of psychological distress (Adikari et al., 2021). The COVID-19 infection continued to exploit the world with over 65 lakh deaths as of October 2022. Social distancing and rigorous enforcements of boundaries with quarantine deeply affected the society causing an increase of panic and anxiety in the general population. Prior works discusses the impact of COVID-19 in increasing the anxiety in people by age groups (Muller et al., 2020).

Cognovi Labs based in the United States is a center for AI-based behavior analysis developed a coronavirus panic index (What Is Cognovi Coronavirus Panic Index and How Does It Work, 2022/2022). The product works on analyzing the emotions of the general population through forums, blogs, social media, and other internet sources especially of a specific area. This further can help the government to analyze the minds of the population and develop rules accordingly. This also guides them to provide mental health awareness in their areas (Top 10 Emotional AI Examples/ Use Cases & Reasons for Success, 2022/2022).

5.3.1 Collection methodology

Fig. 7.6 discusses the collection methodology of a typical panic index. All the data are identified from relationships that are commercial and in public (Cognovi Labs | Alternative Data - Knoema, 2022/2022).

5.3.2 Outcomes

In accordance with the Cognovi Labs and the data collected by them, anger is the most dominant emotion brought by the pandemic. The emotion is followed by fear and sadness. This is even followed by joy which can probably signify faith and hope toward the betterment. As stated by Cognovi Labs, the topic of masks generates the emotion of fear, sadness, and even disgust. Apart from the socio-economic impact the pandemic had, it also led to the postpandemic stress. The panic index can help keep in check the psychological impact that was caused during and after the pandemic (Muller et al., 2020).

5.4 Anura

Hypertension is known to be the leading cause of the majority of cardiovascular diseases. Hypertension is a health issue with almost 1.3 billion people in the world and a cause of death to over 10 million (Hypertension - World Heart Federation, 2022/2022). The accurate and timely measurement of blood pressure hence is significant for both the diagnosis and prognosis of hypertension (Muntner et al., 2019). Dr. Kang Lee, the chief scientist at NuraLogix Corporation working in the American Heart Association with his team, worked upon Anura, a blood pressure measurement technology.

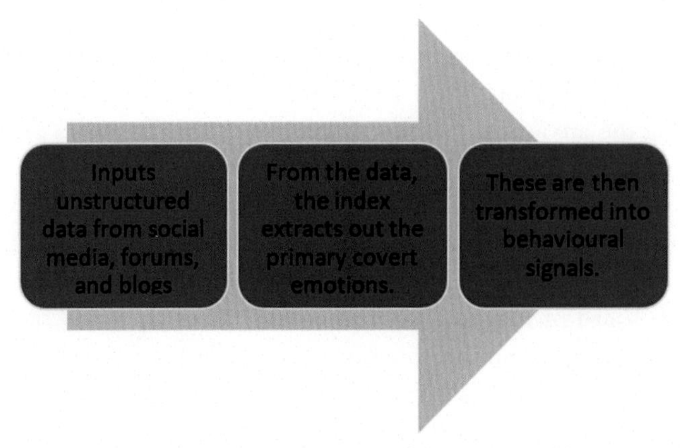

Figure 7.6 The collection methodology of the cognovi lab panic index.

The principle behind is transdermal optical imaging (TOI) (Smartphone to Measure Blood Pressure Research Paper Took Over International News - NuraLogix, 2022/2022). Lee describes his discovery as a happy accident. Lee and Zheng initially were working on building a contactless lie detector. During that study, they figured out how it is similar and can help in detecting blood pressure eventually helping the larger group of the population (Preventative Health at Your Fingertips: U of T Researchers Accurately Measure Blood Pressure Using Phone Camera, 2022/2022).

5.4.1 Principle TOI

Transdermal optical imaging is a relatively novel technology that is used to analyze human stress levels. The technology measures blood flow changes in the face with utmost ease. The task is done without contact and can be performed remotely whenever you want (Wei et al., 2018).

5.4.2 Working of the algorithm

The algorithm behind NuraLogix's Anura is quite complex and based on TOI. It extracts the following features to give accurate results (Top 10 Emotional AI Examples/Use Cases & Reasons for Success, 2022/2022).

1. Physical characteristics including the skin tone, age, and weight.
2. Blood-flow signals through the translucent skin of the human face (the hemoglobin concentration determined by the light reflecting on the skin).

The light foremost reaches the translucent nature of the facial skin and penetrates through reaching the underlying hemoglobin. The red hemoglobin is captured by the optical sensor of the smartphone. This allows the device to observe the facial blood flow changes (Preventative Health at Your Fingertips: U of T Researchers Accurately Measure Blood Pressure Using Phone Camera, 2022/2022). Dr. Kang Lee et al. took sample 2-min videos of 1328 people for their study. The Iphone captures and analyzes the facial blood flow through the translucent skin. As compared to the conventional cuff-based blood pressure measurement machines, it was found that the blood pressure was being measured up to 95% accuracy (Smartphone to Measure Blood Pressure Research Paper Took Over International News - NuraLogix, 2022/2022). The application is widely available in smartphones and can be relied on for an accurate and contactless measurement of blood pressure (NuraLogixTM AnuraTM Contactless Health Vitals Measurement Solution Wins 2021 MedTech Breakthrough Award, 2022/2022). It can achieve this within 30 s (Smartphone to Measure Blood Pressure Research Paper Took Over International News - NuraLogix, 2022/2022). The application was awarded the MedTech Breakthrough award in 2021 for the category of "Best Biometric Solution" (NuraLogixTM AnuraTM Contactless Health Vitals Measurement Solution Wins 2021 MedTech Breakthrough Award, 2022/2022).

5.5 Suicide prediction tools

Suicidal deaths have always been a major concern worldwide (Walsh et al., 2017a). Suicide is undoubtedly one of the most common causes of death in the younger generation (Mishara & Weisstub, 2016). Close to 800,000 people tend to commit suicide every year. The number 800,000 is approximately twice as large as the number of deaths by homicide (Walsh et al., 2017a). People highly prone to suicide often omit engaging with medical professionals leading to the significant danger (Walsh et al., 2017a). Prior works depicts the suicidal rate of top five leading countries with Lesotho being on top followed by Guyana (Suicide Rate by Country 2022, 2022/2022). A lot of research is being conducted and arranged in this area in integration with AI and data science. The research is divided into two significant groups (D'Hotman & Loh, 2020).

1. Medical suicide prediction tools: When the AI tool uses data through government or hospital records to give the outcome, it comes under Medical Suicide Rate Prediction. It is generally done in hospital settings (D'Hotman & Loh, 2020).
2. Social suicide prediction tools: When the AI tool uses data through social media such as Facebook and browsing habits (for e.g., through Google), it comes under social suicide rate prediction. It is generally done in social settings (D'Hotman & Loh, 2020).

In 2018, a pilot project has been announced by The Public Health Agency of Canada. The contract is signed with an Ottawa-based AI company namely Advanced Symbolics, which successfully predicted the presidential results of the Trump election in 2016 and the Brexit referendum. The aim of the project is to highlight the suicide hotspots by identifying suicidal behavior. This would in turn help the government to locate their resources to right areas in right amounts in order to prevent suicidal deaths (D'Hotman & Loh, 2020; Walsh et al., 2017a). Fig. 7.7.

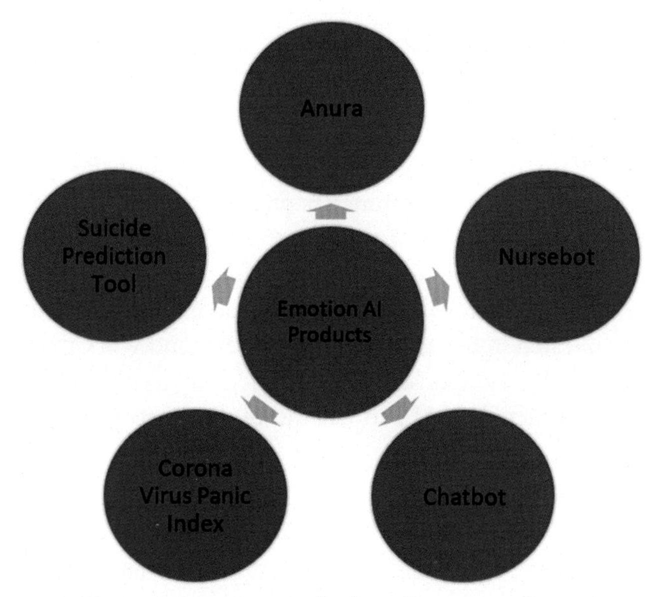

Figure 7.7 Emotion artificial intelligence products.

6. Challenges and issues in EAI and healthcare

Some of the difficulties and challenges faced in and around EAI and healthcare are as follows.

1. Difficulty in integration of EAI healthcare systems during existing healthcare software.
2. Challenges in and around local context and solutions of AI.
3. Challenges around understanding of humans and clarity of the outcomes.
4. Underlying difficulties around trust in AI technologies.

7. Conclusion

The complexity and dynamically changing data in the field reveal that EAI is yet to grow to its fullest in the healthcare domain. A deep exploration can open avenues for leveraging this technology for societal benefits in near future. EAI has the capability to transform the world for good. In fact, overtime, it is believed that certain tasks (for e.g., diagnosis of a disease) are done better by robots than humans. Algorithms are outperforming the specialists at identification, detection, and classification of malignant tumors. Summarily, it can be stated that there is a long way to go to accomplish the production of EAI with sufficiently sophisticated and intelligent algorithms, to attain meaningful and useful outcomes for mankind.

8. Future prospects

Apart from the cost and other such challenges the most significant paradox remains which is the difference between a human and an AI machine. The difference is highlighted when the information provided is equivocal and insufficient. The human brain in such cases often moves ahead to adjust and adapt to the scenario and perform the task, whereas a machine or a robot would mostly fail to do so. This is because the algorithms are still not complex enough and need a lot more research. The algorithms need to be more complicated in terms of decision-making and work more than just simple "yes" and "no" responses. The exponentially growing stream of authors are being attracted to AI and healthcare (Secinaro et al., 2021).

References

Abd-alrazaq, A. A., Alajlani, M., Alalwan, A. A., Bewick, B. M., Gardner, P., & Househ, M. (2019). An overview of the features of chatbots in mental health: A scoping review. *International Journal of Medical Informatics, 132.* https://doi.org/10.1016/j.ijmedinf.2019.103978

Adikari, A., Nawaratne, R., de Silva, D., Ranasinghe, S., Alahakoon, O., & Alahakoon, D. (2021). Emotions of COVID-19: Content analysis of self-reported information using artificial intelligence. *Journal of Medical Internet Research, 23*(4). https://doi.org/10.2196/27341

Ayanouz, S., Abdelhakim, B. A., & Benhmed, M. (2020). A Smart chatbot architecture based NLP and machine learning for health care assistance. In *ACM international conference proceeding series. Association for computing Machinery.* https://doi.org/10.1145/3386723.3387897

Cognovi Labs | alternative data - knoema. (2022). (Original work published 2022).

D'Hotman, D., & Loh, E. (2020). AI enabled suicide prediction tools: A qualitative narrative review. *BMJ Health & Care Informatics, 27*(3). https://doi.org/10.1136/bmjhci-2020-100175

Fazi, M. B. (2019). Can a machine think (anything new)? Automation beyond simulation. *AI & Society, 34*(4), 813–824. https://doi.org/10.1007/s00146-018-0821-0

Fölster, M., Hess, U., & Werheid, K. (2014). Facial age affects emotional expression decoding. *Frontiers in Psychology, 5.* https://doi.org/10.3389/fpsyg.2014.00030

French, R. M. (2000). The turing test: The first 50 years. *Trends in Cognitive Sciences, 4*(3), 115–122. https://doi.org/10.1016/S1364-6613(00)01453-4

How emotion AI can be A game changer in healthcare | Enablex insights. (2022). (Original work published 2022).

Hypertension - world heart federation. (2022). (Original work published 2022).

The importance of emotional AI in business in the 21st Century. (2022). (Original work published 2022).

Machine learning for nursing – 8 current applications | Emerj artificial intelligence research. (2022). (Original work published 2022).

Mental health: Lessons learned in 2020 for 2021 and forward. (2022). (Original work published 2022).

Mishara, B. L., & Weisstub, D. N. (2016). The legal status of suicide: A global review. *International Journal of Law and Psychiatry, 44*, 54–74. https://doi.org/10.1016/j.ijlp.2015.08.032

The most technologies that uses artificial intelligence in USA. (2019). (Original work published 2019).

Muller, A. E., Hafstad, E. V., Himmels, J. P. W., Smedslund, G., Flottorp, S., Stensland, S.Ø., Stroobants, S., Van de Velde, S., & Vist, G. E. (2020). The mental health impact of the covid-19 pandemic on healthcare workers, and interventions to help them: A rapid systematic review. *Psychiatry Research, 293*. https://doi.org/10.1016/j.psychres.2020.113441

Muntner, P., Shimbo, D., Carey, R. M., Charleston, J. B., Gaillard, T., Misra, S., Myers, M. G., Ogedegbe, G., Schwartz, J. E., Townsend, R. R., Urbina, E. M., Viera, A. J., White, W. B., & Wright, J. T. (2019). Measurement of blood pressure in humans: A scientific statement from the American heart association. *Hypertension, 73*(5), E35–E66. https://doi.org/10.1161/HYP.0000000000000087

NuraLogixTM AnuraTM contactless health vitals measurement solution wins 2021 MedTech breakthrough award. (2022). (Original work published 2022).

NurseBot - the index project. (2022). (Original work published 2022).

Oh, K. J., Lee, D., Ko, B., & Choi, H. J. (2017). A chatbot for psychiatric counseling in mental healthcare service based on emotional dialogue analysis and sentence generation. In *Proceedings - 18th IEEE international conference on mobile data management, MDM 2017* (pp. 371–376). Institute of Electrical and Electronics Engineers Inc. https://doi.org/10.1109/MDM.2017.64

Oprescu, A. M., Miró-Amarante, G., García-Díaz, L., Beltrán, L. M., Rey, V. E., & Romero-Ternero, M. (2020). Artificial intelligence in pregnancy: A scoping review. *IEEE Access, 8*, 181450–181484. https://doi.org/10.1109/ACCESS.2020.3028333

Preventative health at your fingertips: U of T researchers accurately measure blood pressure using phone camera. (2022). (Original work published 2022).

Qaraqe, M., Erraguntla, M., & Dave, D. (n.d.). In Lecture Notes in Bioengineering. https://doi.org/10.1007/978-3-030-67303-1_11.

Ranade, A. G., Patel, M., & Magare, A. (2018). Emotion model for artificial intelligence and their applications. In *Pdgc 2018 - 2018 5th international conference on Parallel, distributed and Grid computing* (pp. 335–339). Institute of Electrical and Electronics Engineers Inc. https://doi.org/10.1109/PDGC.2018.8745840

Robotic nurses | computers and robots: Decision-makers in an automated world. (2022). (Original work published 2022).

Secinaro, S., Calandra, D., Secinaro, A., Muthurangu, V., & Biancone, P. (Dec. 2021). The role of artificial intelligence in healthcare: A structured literature review. *BMC Medical Informatics and Decision Making, 21*(1), 1–23. https://doi.org/10.1186/S12911-021-01488-9/FIGURES/12

Smartphone to measure blood pressure research paper took over international news - NuraLogix. (2022). (Original work published 2022).

Suicide rate by country 2022. (2022). (Original work published 2022).

Top 10 emotional AI examples/use cases and reasons for success. (2022). (Original work published 2022).

Walsh, C. G., Ribeiro, J. D., & Franklin, J. C. (2017a). Predicting risk of suicide attempts over time through machine learning. *Clinical Psychological Science, 5*(3), 457–469. https://doi.org/10.1177/2167702617691560

Wei, J., Luo, H., Wu, S. J., Zheng, P. P., Fu, G., & Lee, K. (2018). Transdermal optical imaging reveal basal stress via heart rate variability analysis: A novel methodology comparable to electrocardiography. *Frontiers in Psychology, 9*. https://doi.org/10.3389/fpsyg.2018.00098

What is cognovi coronavirus panic index and how does it work. (2022). (Original work published 2022).

Further reading

Davenport, T., & Kalakota, R. (Jun. 2019). The potential for artificial intelligence in healthcare. *Future Healthcare Journal, 6*(2), 94. https://doi.org/10.7861/FUTUREHOSP.6-2-94

CHAPTER EIGHT

Machine learning model for teaching and emotional intelligence

Mohit Kumar[1], Syam Machinathu Parambil Gangadharan[2] and Nabanita Choudhury[3]

[1]Department of Computer Science & Engineering, Jaypee Institute of Information Technology, Noida, Uttar Pradesh, India
[2]Liverpool John Moores University, Liverpool, United Kingdom
[3]Faculty of Computer Technology, Assam Down Town University, Guwahati, Assam, India

1. Introduction

In an expert civilization where the rate of learning is constantly decreasing efficient skill acquisition and development are critical processes for the success of society. This event signals the beginning of a period in which the employment market will undergo a profound shift, during which new professions and positions will be formed (New Vision for Education, n.d). It is necessary to have excellent training techniques and systems in place to support ongoing, lifelong learning and create viable career and job routes (Misra et al., 2021). E-learning, often known as distance learning, is a kind of learning in which teachers and students are geographically separated from one another but are nevertheless able to communicate with one another via the use of a range of different technologies (Mystakidis et al., 2021). E-learning is often carried out with the assistance of online educational platforms, which make it possible for instructors, students, and material to communicate with one another and engage in cognitive exchanges (Liagkou et al., 2019). These forums have been linking instructors and students from any area of the globe for centuries now, and this has proven more important during hazardous and tough times when social separation has been inescapable (Jan and Vlachopoulos, 2018). E-learning platforms are being used and evolved in tandem with other conventional face-to-face learning techniques by an increasing number of institutions outside of higher education. These institutions are making use of e-learning platforms. The problem here is that not all systems provide good influence designs that may result in a fruitful life

Emotional AI and Human-AI Interactions in Social Networking
ISBN: 978-0-443-19096-4
https://doi.org/10.1016/B978-0-443-19096-4.00014-6

lesson. This is the root of the issue. Platforms of this kind may have issues such as inadequate user involvement, a lack of rich material, or poorly organized or even fully unorganized online learning.

The brain would be unable to prioritize tasks and conclude without emotions. Therefore, emotions are an essential component of cognitive thinking. A person's analytical thinking behavior savior is directed by the facts that they have learned. Early on in a child's educational career, instruction in the fundamental elements that make up emotional intelligence should be a required part of the curriculum (Mystakidis et al., 2019). However, if this were to be introduced as a new subject for students to study, the academic community might object on the grounds that it would place too much of a burden on students and that educators would be unqualified to teach it. Alternately, the findings of this study recommend incorporating the teaching of emotional intelligence into other types of educational endeavors, such as concurrently with the instruction of cognitive abilities (Fragkaki et al., 2022). In addition to this, learning to code makes them adept at mathematics and gives them resilience in the computational realm. These are the kinds of talents that will serve a youngster well throughout his life, no matter what industry he ends up working in. Because it is a well-known fact that children are getting smarter than ever before, but for some reason, their moral values and ethical actions are deteriorating, our plan is to teach children both programming and emotional intelligence at the same time. Both cognitive abilities and emotional intelligence are crucial for success (Anaraki, 2011). There has been a rise in interest in gaining emotional intelligence via interaction with artificial intelligence and robots. When compared to children who were utilizing screen tools such as a smartphone, tablet, or personal computer, children whose play and learning used a robot resulted in much more intense emotional displays (Doukakis and Alexopoulos, 2021). Young students frequently perceive robots as interacting partners and discover that they may teach them new things in the process. In addition, children who do not have a companion encounter several challenges while attempting to improve their abilities and a machine may fill the void. Because research has solely focused on either emotional intelligence or programming, no solutions that combine the two have been investigated. Goal setting, self-awareness, generosity, and empathy are examples of these characteristics. We are talking about emotional attributes like the ability to set and achieve one's own goals, a healthy dose of self-awareness, a caring heart, a kind spirit, and the ability to set and achieve one's own goals.

Clinicians may be able to duplicate the recommended mode of operation and the results gained to inspire creativity in the development of online training systems and procedures (Natsis et al., 2012). A boost in motivation and creativity may improve performance levels. People can be turned from passive recipients to active heroes and executives by utilizing application features that are motivated by the needs of an organization or community. As a result of this research, an algorithmic model that is intended to teach children the principles of computer programming have been proposed. The model has the children control the movements of a robotic system by engaging with a visual programming tool or sketching on a page and then having their work scanned. Alternatively, the model allows the children to control the motion of the robotic system through voice commands. In addition, the trainers can tailor the various components of the workout to the specific needs of their clients. The lessons in the curriculum include very basic aspects of computer programming, such as sequences, Booleans, and loops. In addition, the activities are designed with the tenets of "shot on target," "soul," "socio-cultural," "empathy," and "sharing" as their foundations. The following are the four stages that make up the operation of the system.

1. At the beginning, different scenarios were thought of as obstacles in the form of finding a route and were put into a database of possible scenarios.
2. Once this step has been completed, a teacher will create a one-of-a-kind scenario by making use of the specimens that have been saved in the database. The students will then be tasked with resolving the issue using this scenario as a guide. Because of this, the software has a graphical user interface that was made just for teachers to use.
3. When the student has successfully logged on, the next step is for them to assess the situation and devise a strategy for how to proceed. As a kind of input, the system is willing to accept a movement plan, and this plan can take one of two distinct forms. Using this blueprint, the robot will be given instructions on how to move around the grid. Instead, it is composed of blank paper and symbols that are drawn by hand. The initial format does not involve any kind of contact with the screen in any way. The second method involves the use of a system that is navigated using a keyboard.
4. The robot can determine the pattern of movement by scanning the paper that the student has prepared. Children learn new skills such as looping and traveling ahead, left, or straight to avoid a jump while working on developing a mobility guideline for the robots (the number of strides

in the same direction). The youngsters develop a grasp of the concepts of emotional intelligence known as empathy, sharing, and assisting others, as well as the capacity to apply those concepts in real-world situations. The capacity to understand and share the feelings that are experienced by other people is what is meant by the term.

The methodology that was recommended was successful in teaching children essential programming ideas, thereby strengthening the children's intuitive, emotional, and intellectual capabilities, as well as their moral values. Some of the key contributions made by the study include the following.

1. A computational model is made to teach young children (ages 4–8) about computers and emotional intelligence at the same time. This is done by having them interact with the model for short periods of time.
2. Even though it is rare for students to make mistakes when learning about the different adaptive settings, the model can find and fix any problems that might happen.
3. Researchers have quantified the influence that children's ages have on the way they learn new information. It appears that a child's age had some influence on how they learned.

2. Related work

Capabilities related to emotional intelligence may be strengthened by specialized training to better cope with day-to-day social issues. These capabilities include expressing, comprehending, and controlling one's emotions. Several experiments have been conducted in which emotional intelligence was taught to children in a variety of settings and formats. The authors also found that the method by which learners and tutors analyze and react to emotions may either help or hinder the emotional and social growth of a kid. Educational robots are a fantastic teaching aid for students of any level in today's modern society. Students' ability to learn, from kindergarten on up to college and beyond, is likely to be aided using robots in educational settings. It was discovered that when students created significant dramatic plays with robots, they were able to easily convey their ideas, emotions, and wishes. They asked either a single kid or a group of three youngsters to narrate a narrative to the robots as part of the research study on how children engage with robots. The capacity of children's emotional intelligence was intended to be developed via the employment of socially helpful robots in conjunction with interactive activities. The findings demonstrated that children's emotional intelligence and social abilities increased when they interacted with many robots, whatever the nature of the relationship. Robotics provides a significant new platform for educational endeavors (Garg, 2023). This is made possible by the robot's capacity to communicate with instructors. In these types of settings, a wide range of activities are carried out with the purpose of developing technical abilities such as technical creativity, originality, and the ability to discover solutions to issues. Using robots in the classroom to introduce younger students to the concept of programming and the study of computer science is a growing practice. Block-coding games teach children the fundamentals of computer programming, allowing them to achieve their goal together. Their approach, which employs block programming, has been proven effective in assisting younger pupils in understanding the fundamentals of computational thinking and logic construction. The researcher used pictograms to guide children through the process of controlling and programming a digital robot. According to local legend, Piktomir is the home of a robot named Fidget. Fidget is responsible for restoring the coverings of space shuttles that were damaged during launch. Fidget is torn between moving forward, to the right, and to the left. The kids are the ones guiding the path, ordering everyone else to go, turn right, turn left, and so on. They are also in charge of traffic control. They achieve this by dragging and dropping the relevant pictograms into the appropriate spots on the grid provided by the first technique (Chalki et al., 2019). This is how they approach it. According to the findings of this study, Piktomir is not only useful for developing linear programs, but it may also be useful for creating programs with complex control structures such as loops, conditionals, and subroutines. This is due to Piktomir's capacity as a recursive algorithm to solve problems in a recursive fashion. Piktomir's goal is to make coding entertaining for youngsters by providing the most straightforward and text-free environment possible. The students improved their skills with the route program until they were masters of it. As a result, they were able to comprehend how to program the Dash robot's several alternative pathways. Twenty children aged five to six, along with their preschool teachers, participated in a study on the impact of informal STEM education delivered through coding. The experiment lasted 3 months (in science, technology, engineering, and mathematics). It has been discovered that exposing students to coding using robots in the context of their typical classes helps ignite their interest in the subject and desire to learn it. This was revealed when the students were exposed to coding in the context of their normal classes. KIBO was programmed without the use of displays by connecting several wooden blocks that supply the robot with a variety of commands. With KIBO, you can now do a variety of jobs. The youngsters in the study varied in age from three to 5 years old. Students in this cohort are taught the fundamentals of coding and computational thinking as part of their normal academic program. They elected to use a method known as "code as a playground" as their plan of action. When coding is integrated into several topic areas using a project-based teaching approach, it has been found to be a successful strategy for teaching age-appropriate core programming in primary school courses. This method has been shown to be effective. Engineering, physics, and mathematics are among the subjects covered in this category (Deligiannidi and Howard-Jones, 2015). Elkin and his colleagues created a lesson plan for elementary school children that employs the KIBO robotics kit as the primary means of communicating information about engineering and programming fundamentals. This method was developed for use with primary school students. Computer programming skills can now be taught to children as young as 3 years old. Self-repeating sequences and loops are examples of the basics. This activity would have been difficult to do in the past due to physical limitations.

2.1 Neuroscience and education

The application of cognition to the academic environment is referred to variously as neuropedagogy, pedagogical neurobiology, or neuroeducation. Neuropedagogy is the field that allows for the intersection of science and education, and its primary research objective is to figure out how to activate hitherto unexplored regions of the brain and establish new neural connections. The implementation of neuroscientific principles in learning is not always straightforward, and lecturers in higher education occasionally must deal with many challenges at the same time. Other studies investigate neurobiology illiteracy, most specifically the frequency of incorrect ideas about the brain and learning, which are referred to as cognition, among aspiring and current educators. This study is being done with the intention of assisting academic instructors in improving and expanding their thinking abilities. The most current findings from research on neuroeducation, motivation, and physical exercise have been compiled into a book that focuses on the factors that influence both students and instructors in the field of physical education (Kashyap, 2019a). Another piece of research investigates the pervasiveness of neuromyths in educational settings. According to this piece of writing, even though the field has been the subject of publications for the better part of 2 decades, neuromyths continue to hold sway due to a dearth of scientific understanding and poor-quality information sources. Higher education institutions need to strengthen their scientific curricula to reduce the prevalence of neuromyths in popular culture. Students should use information and communication technology to better their learning, and instructors should utilize these tools to understand how to interact with their students. According to the findings, a pessimistic outlook is often associated with providing responses that are less than accurate while being evaluated.

2.2 Educational neuroscience and online learning

New online platforms are being developed in conjunction with global research programs focusing on educational neuroscience. This trend is occurring as the e-learning industry continues to expand rapidly. The program's findings will be disseminated in the form of a teacher toolkit, a pamphlet titled "Science of Learning Principles," and a free online course for educators dubbed "The Neurology of Education for Instructors" A toolbox for trainers was established as part of this program, which had the added benefit of being a bonus. Neuroandragogy is a type of resistance to the marginalization that many people face. If the goal is to increase the number of underserved individuals who continue their education throughout their lives, the most recent advances in the field of neurodidactics must be incorporated into the adult education process. The outcomes include a teacher education program called "Neuroandragogy in the Education of People from At-Risk Populations," as well as an online learning environment and neurodidactics resources in different languages. Pro-Learn is the name of this useful tool. It is possible to progress toward the goal of lowering the percentage of students who drop out of school before completing their secondary education by fostering children's cognitive abilities and the learning process itself using an integrative methodology and innovative instructional strategies (Doukakis and Alexopoulos, 2020). A further effect of this endeavor to accelerate learning is that fewer students will drop out before completing their degree. The research findings were compiled into a resource pack for educators and teachers that covered topics such as "learner identity," "how to teach kids," "the latest discoveries in the central nervous system," "adapting teaching methods to the needs of students," "memorizing," "being able to work based on the student's funds," "finding the potential of students' brains," "sensorimotor required to teach," and "introductory mentoring to use." This remark necessitates a footnote. One of the deliverables is a resource kit for teachers to use in the classroom. This sentence needs more citation information.

2.3 Design thinking

Approaching difficult challenges with an innovative and adaptable framework is what is known as "design thinking." It employs a human-centered approach to issue resolution via iterative experimentation. The process of idea generation entails going back through several previously completed processes and doing so as often as is required. During the first step of the process, the target group is defined, and data are discovered, gathered, and evaluated in relation to a problem. The data are then analyzed in order to make sense of it and center its meaning on a set of recurring themes that have been identified. During the phase of ideation, concepts are conceived, ranked in order of importance, evaluated, and improved. Following that, these concepts are turned into prototypes for the purpose of experimentation. In the last step, assessment input on prototypes is gathered to keep a record of the lessons learned, move ahead with any weaknesses, address open improvement concerns, and so on. Techniques from the field of creative thinking may be used in the creation of electronic technologies and artifacts by subscribers to encourage more inventiveness and creativity in end users (Tsiara et al., 2019).

3. Cognitive behavior therapy evaluation

The primary goal of the initiative was to educate postsecondary learning professors in novel teaching strategies based on neuroscience findings by using a digital site that is produced for the public of higher education instructors to help them develop longitudinal communication core competencies. It was designed with the applicability of neurology to education in mind, as well as the provision of professional learning opportunities to learners who would be interested in broadening both their understanding and their capabilities in the relevant scientific field. The technique of design thinking is shown in Fig. 8.1. First, a comprehensive data collection effort was conducted with the users, who were the study's focus. The methods and research user data collected from educational academic staff demonstrated a significant level of interest in the topic as well as gaping holes (Ansari et al., 2021). After that, a collection of anecdotes and guides to good resources were put together to form a good practise guide. After

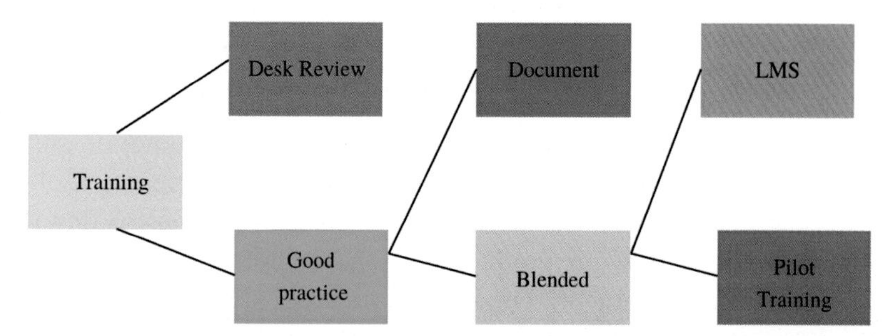

Figure 8.1 The development of a design-thinking-informed online education hub.

that, a thorough brainstorming session was held with the intention of elaborating on the structure of the platform. To begin, a literature review was carried out to locate the most effective software platforms. After that, an online hybrid session of brainstorming and ideation was carried out to identify the platform's ideally suited characteristics and organizational framework. The suggested computational model is made up of four key modules, each of which is designed to educate young students about different programming structures and different emotional intelligence ideas concurrently via a variety of different adaptive situations.

Fig. 8.2 provides the system that will be offered as well as its components. The first part of the course is called "scenario creating," and in it, we develop

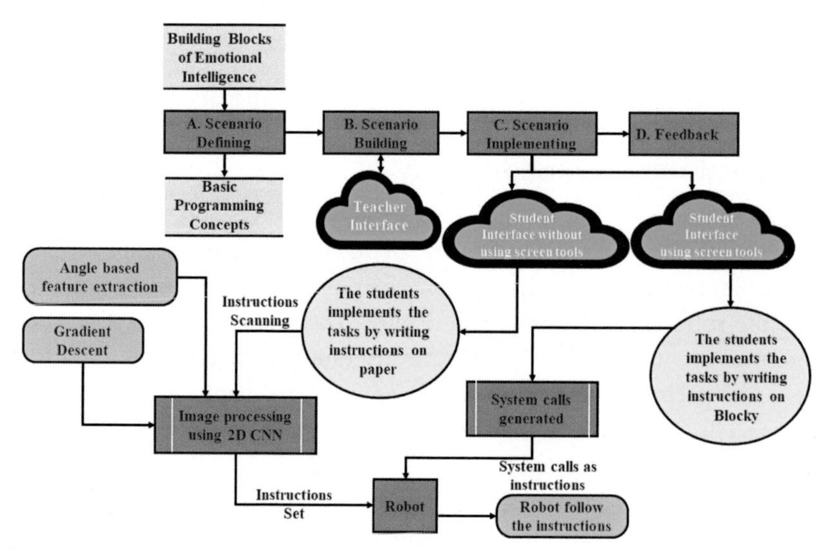

Figure 8.2 A computational framework to meet our goals of training students in both computer science and emotional intelligence.

exercises to identify different routes by using fundamental programming ideas and the fundamental construction blocks of empathy.

The second module is referred to as "scenario construction," and its purpose is for the instructor to construct individualized case studies of route-finding and then assign them without the assistance of a screening tool, as directed by the instructor. Blockly is used to construct the instruction set, and software components are produced for the bot sprite if the learner is using an assessment instrument. The robot does a 2D-CNN analysis on the scanned paper to locate the instructions (Skinner et al., 2019). This is accomplished by the programming of a physical robot known as COZMO. These activities may be designed to take place in parallel. To improve emotional intelligence and essential programming abilities, the set of tasks that have been established is recorded in a database, and they must be carried out in the correct order. The following is an explanation of the actions that serve as examples.

3.1 Introductory activity

The first activity introduces the concept of progressing through stages to achieve a goal. During this portion of the lesson, the child will create a program by making use of a variety of notations. When it determines that it has successfully completed the task, the robot will perform some dancing moves and provide feedback in the form of a "smiley face." It displays a "sad face" as a form of feedback whenever it encounters a problem while carrying out the instructions that have been provided to it (Xiao et al., 2021). A child will get an understanding of the sequence construct of programming as well as the goal-achieving idea of interpersonal skills if they participate in this exercise.

3.2 Self-awareness-based activity

Kids will develop a greater understanding of themselves and their emotions because of participating in this exercise. The objective is to present options to the children so that they may choose their own adventure and learn what causes the robot to feel happy or unhappy. The child devises a plan to go where they want to go by excluding everything that may make them unhappy and packing their journey full of as many happy experiences as possible. The robot obeys the directions written on an interactive board that was specifically developed for it. The child receives education on the programming constructs of sequence and selection (if-else), as well as the self-awareness idea of emotional intelligence.

3.3 Empathy-based activity

Kids are more likely to show compassion after participating in activities focused on empathy. It is beneficial for children to get an understanding of the challenges faced by others. The children plan out a series of maneuvers that the robot will carry out so that it may reach its objective without colliding with any ambulances that may be in its way (Kashyap, 2019b). The ambulance serves as a barrier; thus, the child is required to move out of the path so that the ambulance may pass. A child gains an understanding of the sequencing and selection (if-else) constructs of programming, as well as the empathy idea of emotional intelligence, as they participate in this activity inside an environment that has been specifically created to be interactive.

3.4 Social-awareness-based activity

The concept of social awareness lies at the heart of this activity. The child develops the ability to perceive and respond appropriately to dangerous circumstances in their environment. The automobile is moved in a certain direction by using a concept known as a "loop," which involves traveling numerous blocks in the same direction.

3.5 Sharing-based activity

This practice, which is built on sharing, encourages a child to share his possessions with other people. If a child goes along a road where another, less fortunate child is standing and gives some of his possessions to the disadvantaged child, the child receives a greater reward. This is because four left-hand symbols would be invalid. The students will learn the fundamentals of programming via the repeating (loop) idea that these limitations impose on them. Children acquire an understanding of if-then conditionals when they work to overcome obstacles.

3.6 Scenario building

In the process of learning, we think each pupil should be provided with an environment that can be modified to accommodate his or her unique learning characteristics. These flexible educational opportunities inspire both an extrinsic and an intrinsic sense of desire in the students who participate. In the development of screen tools, the minimalist API Flask, which is based on the computer program Python, is utilized. It works as a framework for web applications and makes calls to the robots or their sprites so that they may run and carry out the command. Fig. 8.3 provides a visual representation of the structural movement of the Screen Tool's user interface.

The entire topology of the deep convolution network is shown in Fig. 8.4. An essential part of both image processing and picture classification, the 2D convolutional neural network (CNN) technique was used. It can identify a wide range of objects and extrapolate a depth estimate from the pictures. Two-dimensional CNNs may derive characteristics from raw image data. We used data from over 4000 images taken in a variety of directions to train the CNN model. Eighty percent of these are used for training, and 20 percent are put to the test.

The 2D convolutional neural network was trained using the 80×80 photo dataset, which resulted in the formation of eight convolutional layers. These intermediate levels are distinguished by the fully connected (FC) layer, the max-pooling layer, and the higher performance of soft-max, as shown in Fig. 8.5.

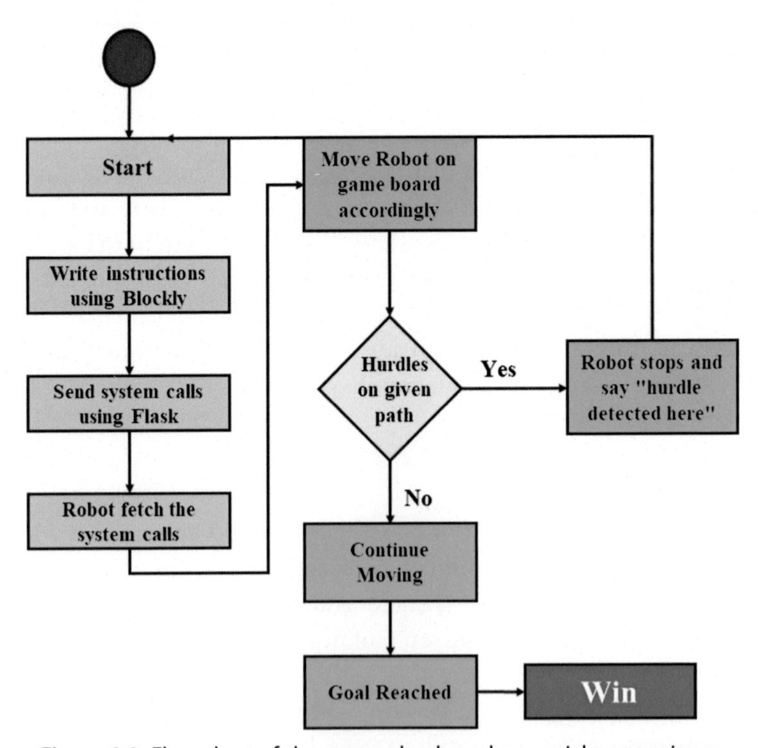

Figure 8.3 Flow chart of the game plan based on social neuroscience.

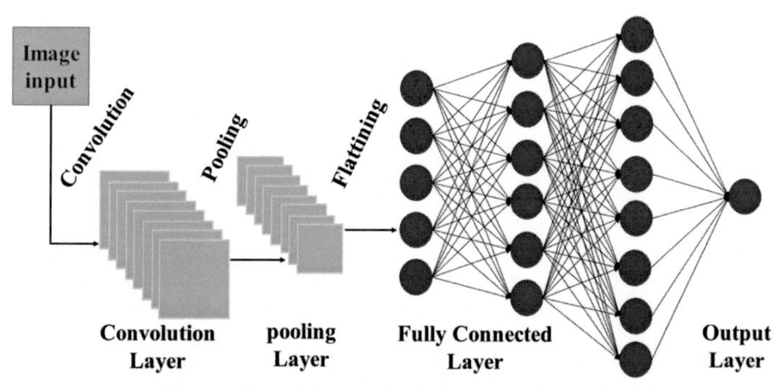

Figure 8.4 The standard CNN architecture.

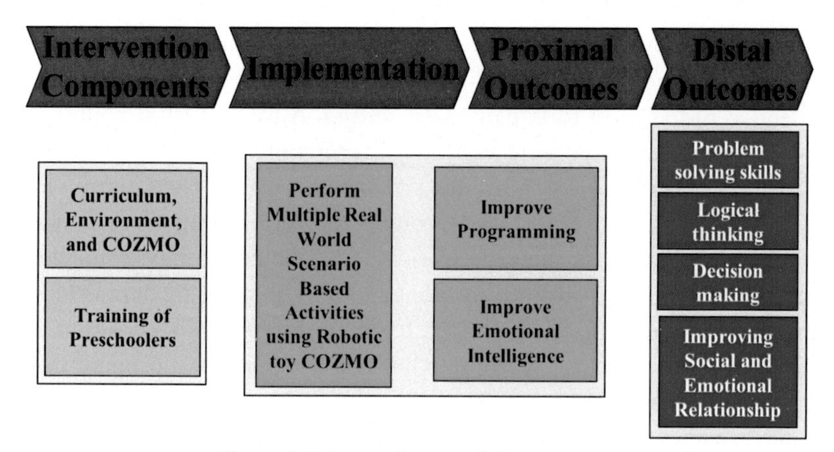

Figure 8.5 Internal steps of the activities.

Gradient descent is an optimization approach that is used to discover a solution rapidly and accurately. Photographic qualities were abstracted in this study by classifying them according to their distinct axes of rotation. Even though less than 500 iterations at a speed of 0.001 were required, the training was finished in under a minute. The success rate was 95% due to the general effectiveness of the talk.

Despite the lack of a screening program, a flowchart explaining the procedure for administering the death penalty via injection was developed. TensorFlow, a deep learning toolkit, was used to retrain the DNNs used in this study so that they could be used for inference. TensorFlow was used to decode the data, which benefited from the child's visual annotations of the various indications.

3.7 Feedback

The program instructions are decoded by the robot, which then operates in accordance with those instructions. After the participant has finished the exercise that was given to them, the review module is going to provide them with positive feedback based on how well they performed in the activity. As a kind of feedback, the robot will communicate its accomplishment by creating ecstatic emotions and noises of excitement after it has arrived at its target. If the robot is unable to reach its target because it was impeded in any way along its journey, it displays a sorrowful countenance and emits alarming noises.

4. Experimental study

Our proposed computational framework was put to the test in an experimental study with school-aged youngsters. Children as young as four and as old as eight took part in the testing. A total of 8500 youngsters across five sample groups were analyzed. Classes were divided up according to the students' ages. Group A had the youngest average age at 4, Group B at 5, Group C at 6, Group D at 7, and Group E at 8. We set up a lab and taught the kids how to perform the experiment so that we could draw some conclusions. Because of this, robotic motions and activities that are connected to them may now be performed independently of our methods, thanks to the layered approach that we have adopted. Using an application programming interface (API), our system might be linked with any robot, provided that the latter had all the necessary fundamental features and motions. The COZMO may be replaced with any other robot that has all the functions of motion (such as left, right, up, and down) and a camera that can be used for scanning purposes (Nair et al., 2021). However, for the robot to successfully carry out the instructions, it must first be linked to our system through API calls. The recommended technique is to be thanked for the remarkable success that was achieved in the experiments. The instructional web platform that needed to be created was going to include two different hybrid subsystems. The first, a learning management system, is geared toward higher education lecturers and is designed to accommodate all the newly produced training content. This technique of instruction is based on neuroscience. In the second area, academic faculty members were able to communicate with one another, share resources, and engage in mutual and peer-to-peer learning, which served as a cooperative way to create a virtual community of inquiry.

WordPress is a widely used and expansive software for managing content that has a significant number of users, installations, and a broad range of plugins and expansions. WordPress is also available for free. Additionally, WordPress offers a larger opportunity for social interaction. WordPress could be utilized for the online discussion board, while Moodle will most likely be used for lectures and other online content. Both systems are open source. These two systems can be used interchangeably because they are comparable (Kashyap et al., 2022). Moodle, a free and open-source online learning platform, has been used by educators all around the world. Despite this, they still struggled to interact with each other. As a result, developing a single sign-on system has become one of the most crucial responsibilities. With the implementation of this technique, users would only need to register once to gain access to the e-learning courses and the platform for communities of practice. This system's users would have access to both the system and the online learning courses. There will come a day in the not-too-distant future when resolving this problem will be critical. The second part of our discussion focused on the aspects of the learning platform and community that we found most important and appealing. The criteria for this hybrid system were created through a brainstorming session separated into two distinct parts. The first step was to create a spreadsheet in Google Drive so that the team could cooperate on the procedure. People involved in the creation of the educational platform, as well as the community at large, gave their thoughts on this page about the elements they thought were most important in terms of the overall image. The following are some examples of sections and columns present in the spreadsheet: Some of the characteristics to consider are aesthetics, communication, content, functionality, category, and criticality (critical, essential, good to have, avoid). Ensure that your work contains at least the following items in that order: having a tenuous connection to the study of neurology (either the theory, the technology, or the methodology), An Analysis of Its Importance to the Neurosciences NP Reference/s, with an Illustration, Screenshot, or Online Resource Link to the Project Reference (theory, approach, or technique), despite the fact that each pair kept their own sheet, all of the participants had unrestricted access to the master sheet and could read the data provided by the other pairs (Mohanakurup et al., 2022). A platform that is already in use, the instructional design for an e-learning module that is now being built, and the use of various neuropedagogical concepts are some instances of potential evidence that supports the hypothesis.

Everyone who took part in the second phase, an online workshop, used an application called Miro to discuss and put together the ideas provided during the first round. We moved on to the next stage after completing the previous one. To ensure that no one entered the discussion with any previous notions about the topics being discussed, each spouse then gave a quick overview of their ideas and suggestions. Second, each concept was written down on a digital post-it note before being placed in the two-dimensional space (Nair et al., 2020). As soon as they arrived, they were sorted into numerous categories depending on the communication, substance, usefulness, and aesthetics of the given solution. Each of these platforms' learning management systems as well as their community-based subsystems has all been changed to reflect this change (Fig. 8.6). Then, for both sets of criteria, we tallied all the entries, read through each one, and analyzed them to determine if there were any substantial inconsistencies or conflicting perspectives on any of the research themes. A consensus was achieved because of this process, and it is that consensus that will be discussed in the next section. In conclusion, supplementary notes were compiled each time concerns with requirements or unanswered questions were discovered. With the help of these open-ended questions, the conversation was able to go to deeper levels and establish a more basic connection with the project's logic. However, the issue of WordPress and Moodle

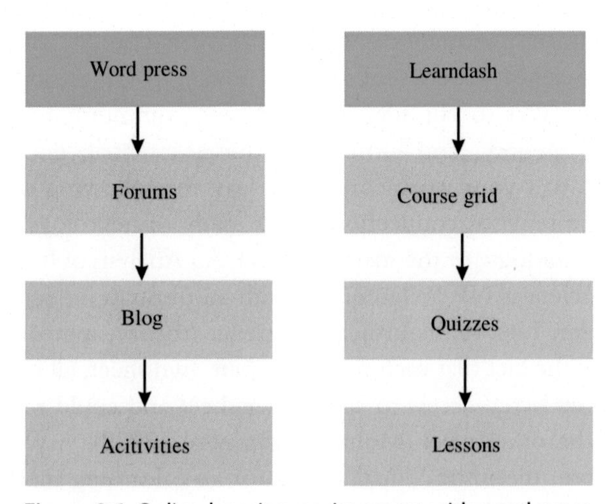

Figure 8.6 Online learning environment with word press.

interacting with one another in a smooth manner was not yet resolved. Table 8.1: A Contrast Between the Subject's Emotional Intelligence Levels Before and After Participating in the Experiment.

The findings indicate that the degree of emotional intelligence has increased across the board for all age categories. However, when compared to younger age groups, older children developed significantly more emotional intelligence.

In addition, we carried out separate analyses on how girls and boys of the same age performed, as shown in Table 8.2 and Fig. 8.7. It was shown that there is not a significant gap in the emotional maturity of males and females of the same age between the ages of 4 and 8, when they are both considered to be extremely young. Table 8.3 shows the results of an emotional intelligence test that was given to both boys and girls.

The comparison findings are detailed in Fig. 8.8 and Fig. 8.9 and Table 8.3 which may be seen below.

5. Discussion

According to the findings of the study, it is beneficial to instruct younger children in the fundamentals of programming in an atmosphere that is conducive to interactive learning, adequate training, and well-designed curricula. Alongside this, the recommended course of study has helped to develop the students' emotional intelligence. The instructional material that was used in this investigation placed an emphasis on fundamental aspects of programming, such as sequencing, iterations (loops), and choosing (conditions). Furthermore, it emphasized a person's emotional intelligence as well as moral ideals. Using the instructions that were based on arrows, the preschoolers who participated in the research were able to

Table 8.1 Sums up the findings of a study that compared the participants' emotional and intellectual capacities.

Group names	Mean ages of the participants (in years)	EI score (out of 20)	
		Before	After
A	5.3	8.5	9.3
B	6.2	8.8	9.9
C	7.5	9.4	11.3
D	8.4	9.7	12.0
E	9.3	10.2	13.2

Table 8.2 The evaluation of the enhancement of emotional intelligence.

Groups	Ages of the participants	EI score of boys (out of 20)	EI score of girls (out of 20)
		Before	After
A	5.3	9.3	9.2
B	6.2	9.8	9.8
C	7.5	11.3	11.3
D	8.4	11.9	12.0
E	9.3	13.2	13.3

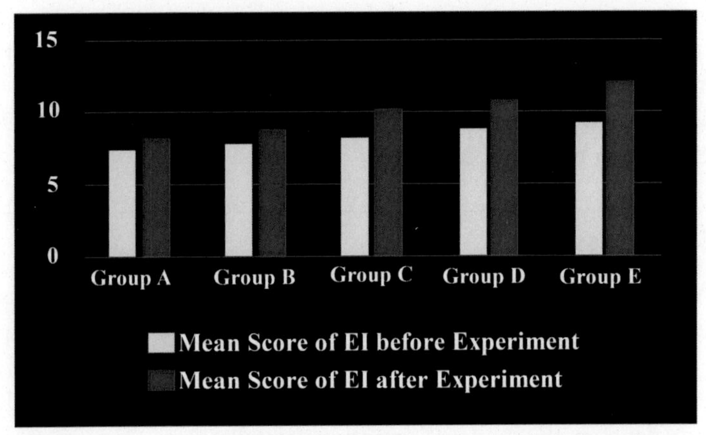

Figure 8.7 The results of each group's programming instructions compare to each other.

Table 8.3 An analysis of the differences in the way boys and girls learn from programming.

Groups	Ages of the participants	Boys learning (out of 20)	Girls learning (out of 20)
A	5.3	11.7	11.6
B	6.2	13.6	13.6
C	7.5	14.5	14.6
D	8.4	16.3	16.2
E	9.3	18.2	18.2

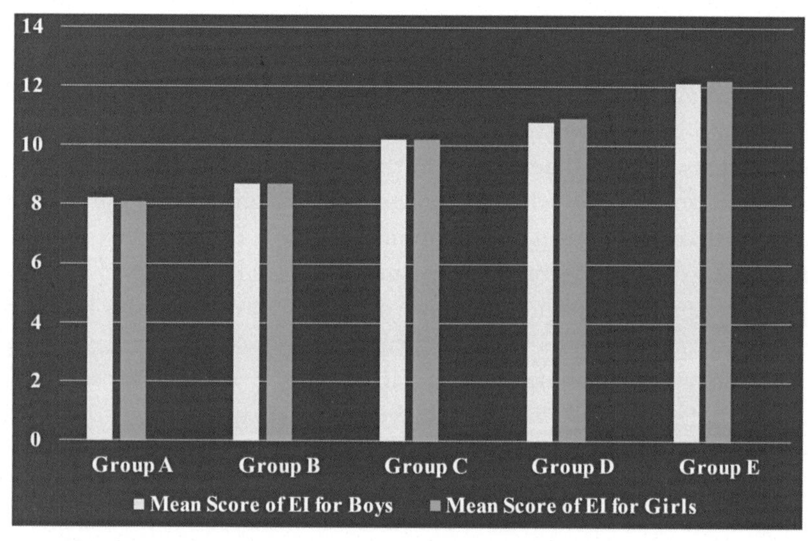

Figure 8.8 The differences and similarities between men and women.

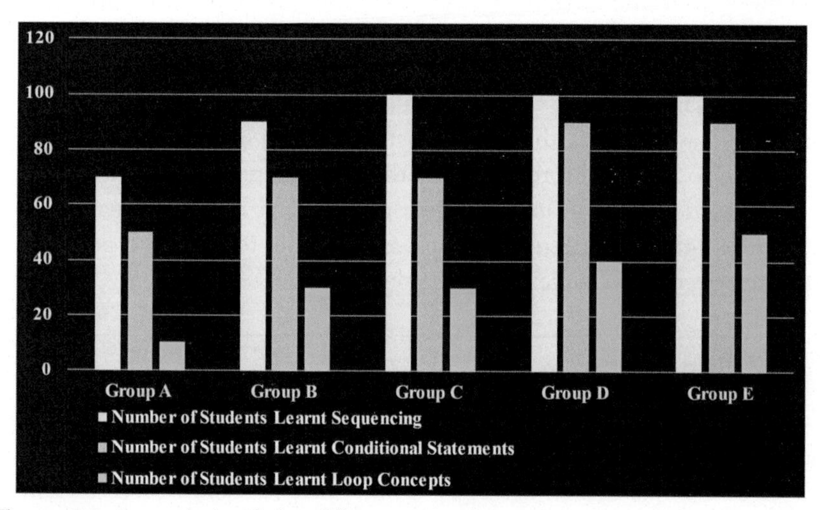

Figure 8.9 An analysis of the differences in the way boys and girls learn from programming.

effectively construct a program that was logically sound. It was made feasible through engaging training sessions that fostered a sense of curiosity in the young participants. Younger pupils, those aged 4 and 5, were slower in their performance of these exercises than the older students, who were able to effectively generate conditional assertions. Based on these findings, it seems that the notion of loops in programming may be somewhat difficult for novice programmers to grasp. The kinds of mistakes that students made show that most of them did not have a good understanding of the syntactical rules that go into making loops. There is a consistent finding across all the findings that indicates older children (those aged six and above) displayed a greater mean degree of advancement across the board about programming principles than did younger children (under 6 years). The key causes include a much larger working memory, a greater concentration level, and an expanded capacity to plan. The scope of this study encompasses a variety of programming principles with the goal of assisting children in developing their capacity for reasoning and analysis. We made available an adaptable teacher interface module, which enables educators to generate and distribute student-specific, individualized learning situations in accordance with the pedagogical learning viewpoint. The development of youngsters' emotional intelligence may also benefit from our research, in addition to their ability to master the fundamentals of programming. This progress was made possible because of the well-thought-out exercises that emphasized the significance of developing one's moral principles and character. The youngsters' emotional intelligence improves because they participate in these activities. The findings indicate that engaging children in productive and creative activities, such as playing with robots, is beneficial. In a similar vein, the ability of children to learn is improved when they are given constructive criticism or suggestions on how to improve their performance and avoid making the same errors again. We decided that it was important to give both praise and constructive criticism to get better results.

6. Conclusion

This essay will go through the experimental approach that was used as well as the building of a dual, community-based e-learning platform. We shall do so with an eye on the opportunities that educational neuroscience brings. At this moment, the platform's foundation was being erected just here. Clinicians may be able to duplicate the recommended mode of operation, and the results gained to inspire creativity in the development of online training systems and procedures. Furthermore, physicians can change the outcomes. This will expand the number of students who can study online. To use this strategy, you must avoid the typical learning management system, which limits online learning to a fixed collection of resources, examinations, and homework. People can be turned from passive recipients to active heroes and executives by utilizing application features that are motivated by the needs of an organization or community. This change might occur in a matter of seconds. A boost in motivation and creativity may improve performance levels and education. According to the findings of this study, increasing calculations are based on mathematical computation to achieve the goal of educating children in programming concepts while also improving their emotional intelligence. This is because there is a dearth of educational research that has been shown to be effective. The findings of the research have shown that it is feasible to educate younger children in the principles of programming as well as emotional intelligence without requiring them to actively engage with a screen. Additionally, it has been determined that the use of robots results in increased levels of participation and performance on the part of pupils. The cost of automation is the fundamental restriction that must be overcome, and this expense must be reduced as much as possible by manufacturing robots domestically rather than importing them.

References

Anaraki, F. B. (2011). Elearning and mLearning at assumption university. In *Proceeding of the international conference on E-education, entertainment and E-management*. https://doi.org/10.1109/iceeem.2011.6137832

Ansari, G., Garg, M., & Saxena, C. (December 2021). Data augmentation for mental health classification on social media. In *Proceedings of the 18th international conference on natural language processing (ICON)* (pp. 152–161).

Chalki, P., Mikropoulos, T. A., & Tsiara, A. (2019). A delphi study on the design of digital educational games. In *Lecture notes in computer science* (pp. 433–444). https://doi.org/10.1007/978-3-030-23560-4_32

Deligiannidi, K., & Howard-Jones, P. (2015). The neuroscience literacy of teachers in Greece. *Procedia - Social and Behavioral Sciences, 174*, 3909–3915. https://doi.org/10.1016/j.sbspro.2015.01.1133

Doukakis, S., & Alexopoulos, E. C. (2021). The role of educational neuroscience in distance learning. knowledge transformation opportunities. In *Advances in intelligent systems and computing* (pp. 159–168). https://doi.org/10.1007/978-3-030-67209-6_18

Doukakis, S., & Alexopoulos, E. C. (2020). *Knowledge transformation and distance learning for secondary education students - the role of educational neuroscience. 2020 5th south-east Europe design automation, computer engineering, computer networks and social media conference.* SEEDA-CECNSM. https://doi.org/10.1109/seeda-cecnsm49515.2020.9221821

Fragkaki, M., Mystakidis, S., & Dimitropoulos, K. (2022). Higher education faculty perceptions and needs on neuroeducation in teaching and learning. *Education Sciences, 12*(10), 707. https://doi.org/10.3390/educsci12100707

Garg, M. (2023). Mental health analysis in social media posts: A survey. *Archives of Computational Methods in Engineering, 30*, 1819–1842. https://doi.org/10.1007/s11831-022-09863-z

Jan, S. K., & Vlachopoulos, P. (2018). Influence of learning design of the formation of online communities of learning. *International Review of Research in Open and Distance Learning, 19*(4). https://doi.org/10.19173/irrodl.v19i4.3620

Kashyap, R. (2019a). Security, reliability, and performance assessment for healthcare biometrics. In *Advances in medical technologies and clinical practice* (pp. 29—54). https://doi.org/10.4018/978-1-5225-7525-2.ch002

Kashyap, R. (2019b). Big data analytics challenges and solutions. In *Big data analytics for intelligent healthcare management* (pp. 19—41). https://doi.org/10.1016/b978-0-12-818146-1.00002-7

Kashyap, R., Nair, R., Gangadharan, S. M., Botto-Tobar, M., Farooq, S., & Rizwan, A. (2022). Glaucoma detection and classification using improved U-Net Deep Learning Model. *Healthcare, 10*(12), 2497. https://doi.org/10.3390/healthcare10122497

Liagkou, V., Salmas, D., & Stylios, C. (2019). Realizing virtual reality learning environment for industry 4.0. *Procedia CIRP, 79,* 712—717. https://doi.org/10.1016/j.procir.2019.02.025

Misra, S., Roy, C., & Mukherjee, A. (2021). *Introduction to industrial internet of things and industry 4.0.* https://doi.org/10.1201/9781003020905

Mohanakurup, V., Parambil Gangadharan, S. M., Goel, P., Verma, D., Alshehri, S., Kashyap, R., & Malakhil, B. (2022). Breast cancer detection on histopathological images using a composite dilated Backbone Network. *Computational Intelligence and Neuroscience, 2022,* 1—10. https://doi.org/10.1155/2022/8517706

Mystakidis, S., Berki, E., & Valtanen, J. (2019). The Patras blended strategy model for deep and meaningful learning in quality life-long distance education. *Electronic Journal of e-Learning, 17*(2). https://doi.org/10.34190/jel.17.2.01

Mystakidis, S., Papantzikos, G., & Stylios, C. (2021). *Virtual reality escape rooms for STEM education in industry 4.0: Greek teachers perspectives. 2021 6th south-east Europe design automation, computer engineering, computer networks and social media conference.* SEEDA-CECNSM. https://doi.org/10.1109/seeda-cecnsm53056.2021.9566265

Nair, R., Singh, D. K., Ashu, Yadav, S., & Bakshi, S. (2020). Hand gesture recognition system for physically challenged people using IOT. In *2020 6th international conference on advanced computing and communication systems (ICACCS).* https://doi.org/10.1109/icaccs48705.2020.9074226

Nair, R., Vishwakarma, S., Soni, M., Patel, T., & Joshi, S. (2021). Detection of covid-19 cases through X-ray images using hybrid deep neural network. *World Journal of Engineering, 19*(1), 33—39. https://doi.org/10.1108/wje-10-2020-0529

Natsis, A., Vrellis, I., Papachristos, N. M., & Mikropoulos, T. A. (2012). Technological factors, user characteristics and didactic strategies in educational virtual environments. In *2012 IEEE 12th international conference on advanced learning technologies.* https://doi.org/10.1109/icalt.2012.67

New Vision for Education - World Economic Forum. (n.d.). Retrieved February 27, 2023, from https://www3.weforum.org/docs/WEFUSA_NewVisionforEducation_Report2015.pdf.

Skinner, D., Kendall, H., Skinner, H. M., & Campbell, C. (2019). Mental health simulation: Effects on students' anxiety and examination scores. *Clinical Simulation in Nursing, 35,* 33—37. https://doi.org/10.1016/j.ecns.2019.06.002

Tsiara, A., Mikropoulos, T. A., & Chalki, P. (2019). EEG systems for educational neuroscience. Universal access in human-computer interaction. In *Multimodality and assistive environments* (pp. 575—586). https://doi.org/10.1007/978-3-030-23563-5_45

Xiao, R., Zhang, C., Lai, Q., Hou, Y., & Zhang, X. (2021). Applicability of the dual-factor model of mental health in the mental health screening of Chinese College Students. *Frontiers in Psychology, 11.* https://doi.org/10.3389/fpsyg.2020.549036

CHAPTER NINE

Emotion AI: Cognitive behavioral therapy for teens having some mental health disorders

Mohammed Hasan Ali Al-Abyadh[1,2] and Vinh Truong Hoang[3]
[1]Department of Special Education, College of Education in Wadi Alddawasir, Prince Sattam bin Abdulaziz University, Saudi Arabia
[2]College of Education, Thamar University, Dhamar, Yemen
[3]Faculty of Computer Science, Ho Chi Minh City Open University, Ho Chi Minh City, Vietnam

1. Introduction

Anxiety disorders are a common kind of illness in youngsters, and according to research, the family and community may have a role in children's and teens' mood and anxiety difficulties. Academic incapacity is likely to deteriorate without treatment, and as children get older, they are more susceptible to anxiety disorders (Mannarini et al., 2016). Many problems need early intervention to be successfully treated. This is becoming more common in all age groups. Substantial advances in the treatment of these challenges have been made because of significant developments in the application of cognitive-behavioral therapy, which has resulted in considerable progress. All these elements remain the same whether the therapy is given to a single person or a complete family, whether it is used alone or in conjunction with another medicine, such as sertraline, and whether it is provided as a stand-alone treatment or in conjunction with other therapies. At least eight pilot studies have shown that in vivo desensitization, video, live, and participatory modeling are beneficial in the treatment of anxiety and specific phobias in children as young as preschool and kindergarten (Song et al., 2021). Graded exposure, graded exposure with reward, and graded exposure with participatory modeling are among these strategies. They also use video, live modeling, and participatory modeling. These medicines have been demonstrated to help reduce the worried and apprehensive feelings that are common in children of this age. These treatments were administered to children who had previously displayed symptoms of anxiety disorders or phobias. As a result, certain age groups are more amenable to the

Emotional AI and Human-AI Interactions in Social Networking
ISBN: 978-0-443-19096-4
https://doi.org/10.1016/B978-0-443-19096-4.00001-8

idea of trying treatment. Despite this, the youngest children were not adequately represented in a couple of the studies, as the mean ages of the samples were 11.03 and 7.8, respectively. This is because people of a certain age are more open to the concept of trying treatment.

Other studies (Guo et al., 2020) included children as young as 5, but no subgroup analysis was performed. While some studies included only very young children or no children under the age of five, others included children as young as five. During a study conducted in the 1950 and 1960s, many patients with panic disorder either improved or showed no change. Prior research was conducted after anxiety was successfully identified. Medication yielded contradictory findings (Guo et al., 2020). Individuals did not substantially improve with time, according to several of these studies. Furthermore, some people believe that young children are not mature enough to benefit from cognitive behavior therapy (CBT). This might be one of the reasons why studies using CBT approaches to treat severe childhood anxiety disorders did not involve enough younger children. Recent research, however, has called these assumptions into question. To begin with, research has revealed that toddlers are equally as likely as older children to suffer from persistent anxiety. According to factor-analytic research, symptom presentations in preschoolers are comparable to those in older children (Moser et al., 2008). Existing research indicates that teenagers' social ties and interpersonal skills have a major influence on their mental health (Ran et al., 2022). For junior high school adolescents, puberty begins, which is a vital moment in their lives. Social anxiety may be examined in a variety of ways in several industrialized nations. The incidence and prevalence rates for each of their respective categories are significantly higher than the national average. Adolescents are the age group most prone to experiencing depression, which is a big component of children's overall poor mental health (ALHarbi et al., 2022) and has a considerable influence on this condition. Adolescents are profoundly impacted, particularly in areas vital to their everyday lives, educational aspirations, and development potential. The study's two goals are to (1) contribute to the field's understanding of how to lessen social anxiety in adolescents and (2) have an impact on current initiatives and programs for mental health education. Please let them know if you have any experience assisting teenagers in overcoming social anxiety.

So, as it goes deeper into the topic of social anxiety in teens, their social skills, communication skills, and ability to adapt to new environments will all get better. The following are the chapter's distinguishing features: In this part, we provide a framework for identifying and assessing the efficacy of cognitive behavioral treatment for adolescent social anxiety. Following that, we constructed a model to predict whether cognitive behavior therapy will have a positive or negative influence. A multiobjective evolutionary algorithm for teenage social anxiety We also constructed a correlation-prediction model to see how well CBT works in treating social anxiety in teens. We assessed the quality and originality of our work by comparing it to current prediction models (Cao et al., 2022). The suggested technique may successfully reduce social anxiety in teens while also improving their interpersonal skills. Our technique is much more intricate and accurate than past studies, and it is also far more effective in reducing teens' social anxiety.

2. Related work

The mental health issues that today's children face have increasingly captured the attention of society. Numerous studies on adolescent social anxiety have shown promising outcomes. The study's findings on social anxiety and sorrow might be used to promote healthy emotional development in addition to the students' mental health (Xiao & AF, 2021/2021). Profile analysis and logistic regression were used to investigate the relevant data. Even after controlling for demographic characteristics, depressive symptoms and social anxiety were shown to be inversely related to college students who were skilled at expressing and discriminating their emotions. According to the data, alexithymia among college students may be classified in many ways. Outside of the DSM-V, which only mentions social anxiety and depressive symptoms, there are other possible groupings. Colleges that implement well-planned, targeted programs to increase students' capacity to regulate their emotions may be able to greatly improve their students' mental health (RAN et al., 2022). However, it is important to note that using this approach to assist university students suffering from social anxiety has not been shown to be effective. This study advances our understanding of how mobile social media metrics affect adolescent mental health by first investigating the causes of the phenomenon and then developing a model to explain it. It contributes to the advancement of psychology by expanding our understanding of adolescent behavior and providing new advice for assisting people who are suffering from the psychological effects of excessive social media use (Laczkovics et al., 2021). It will also investigate the interaction-driven relationships that exist between various components and identify the linkages that exist between them. To accomplish this, we employ a method that examines the elements of the teen social media user model from four distinct perspectives. These interventions employ cognitive-behavioral therapy and are frequently delivered via social media.

By giving teens a platform and service for managing their social media that is based on evidence, social anxiety has not gone down (Swee et al., 2021, pp. 93—114). This study on social media platforms that emphasize user-generated content focuses primarily on adolescents. The relevant cognitive psychology theories were used to develop the theoretical research model for this area of study. The model's context is teenagers' use of social media and information behavior, with an emphasis on the generational aspects the material itself presents. The distribution of questionnaires allows for the necessary data gathering, which is subsequently used to assess the model. According to the findings of this research, the quality of the content and services has a beneficial influence on the adolescents' anticipation of confirmation. The efficacy of the system, on the other hand, will have little bearing on whether children's expectations are satisfied. Young people's attitudes toward usage of social media influence how they react to stress. However, since individual quality is psychologically malleable, setting limits on intensity is straightforward. As a result, teens' social skills do not increase (Waldron et al., 2018). Socially anxious teenage patients showed improvements in their ability to control their emotions after having cognitive behavior therapy in a group environment. Over a 12-week period, 61 anxious youths were randomly assigned to one of three cognitive-behavioral group therapy (CBGT) groups. The therapy group included 32 individuals, whereas the untreated control group had 29. These two scales assess a person's ability to control their emotions. After dividing the patients into two groups, scores were tallied for both groups of patients who had been diagnosed with social anxiety in adolescents. Adolescents in the CBGT group reported significantly higher levels of self-efficacy in emotion regulation after treatment when compared to their levels prior to treatment. Furthermore, their social anxiety test scores were significantly lower. When the control group and the 12-week treatment group were compared, there was no discernible difference. According to conventional wisdom, given that research, CBGT may help teenagers with social anxiety and enhance their emotional control (Gómez et al., 2017). Social anxiety disorder is the third most common mental disorder. Where it spreads in all societies and may be more prevalent in societies that place great importance on people's opinion, and it is one of the disturbances that disrupt a person's relationship with others, as it affects the individual's production in academic and practical life, in addition to the many psychological effects and psychological exhaustion, and its great impact on society through the individual avoiding social situations, or resorting to alcohol abuse, and it may affect the family life of the individual in the future (Al-Abyadh, 2010/2010).

3. Cognitive behavior therapy evaluation

As a result, socially anxious teenagers can now be identified and treated appropriately. We will be able to assess young people who suffer from social anxiety in a more objective and scientific manner with the help of this tool. Look at Fig. 9.1 to learn more about the B/S mode assessment system, also known as "browser-server mode." Fig. 9.1 shows an example of a client browser that performs fewer logical activities. As a result, the client, browser, and server build a three-tier structural framework to support the system's activities. This model's assessment approach may lower the expenses involved with designing, updating, and bug-fixing it. For a reasonable fee, the client's task is simplified, and the system's performance is improved (Keil et al., 2022).

Teenagers who are socially apprehensive use their browser to submit a server request. When teenage users make requests for media such as images or movies, the browser shows the requested item as it downloads (Şimşek et al., 2020). The database server then receives the server's request and gathers the necessary information for processing before providing the processed data to the client machine's browser (Vincent et al., 2020).

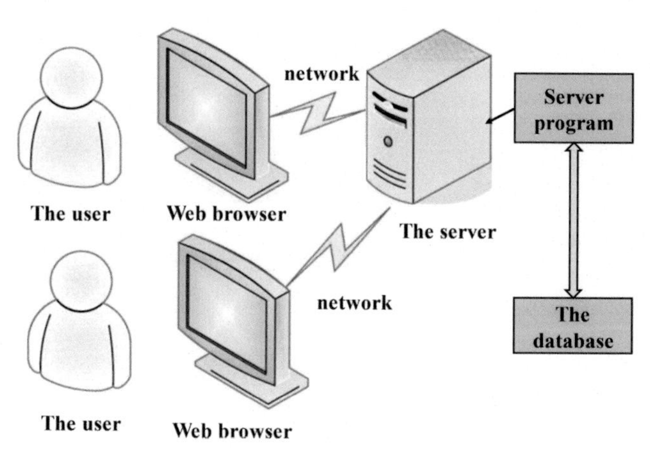

Figure 9.1 The browser and server architecture.

3.1 Development of a cognitive-behavioral assessment for adolescent social anxiety disorder

Diagnosis and Treatment Evaluation. This evaluation is used to diagnose adolescent social anxiety disorder in addition to measuring cognitive and behavioral symptoms. The therapeutic assessment approach for adolescent social anxiety is made up of three parts. These modules serve to validate certification and quality and sort physicians into appropriate groups based on their official job titles. The responsibilities of chief medical officers and staff physicians, as well as constraints on who may do specific types of surgery professional medical professionals have more discretion while executing surgical procedures than their less-experienced peers. Its purpose is to assess whether you are qualified for anything. The cognitive behavior diagnostic and assessment system module for teenage social anxiety is shown in Fig. 9.2 's functional design.

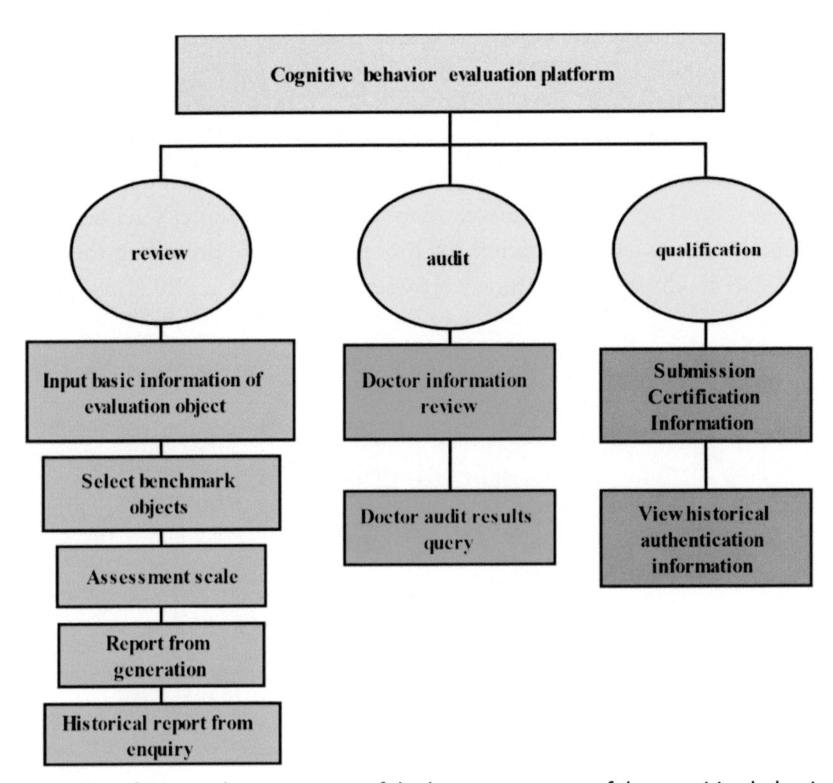

Figure 9.2 A functional arrangement of the key components of the cognitive behavior assessment platform.

3.1.1 Login and registration procedures for young users

After completing the registration process and receiving permission, new users will have access to the cognitive behavior diagnostic and therapy assessment system. Before granting access to the assessment page, the chief medical officer verifies that the user's submitted information is correct and validates it against a database (Mills & Strawn, 2020). To log in, you will need either a username or a mobile phone number. To log in after registering, you must use the username and password you chose during registration must be typed into a login page to have access to the cloud-based diagnostic and therapy assessment platform. Even if a user forgets their password, they can still access the system (Fitzgerald et al., 2019).

3.1.2 A tool for user evaluation

Newly enrolled and authorized users get access to the assessment features. Users (doctors) who have enrolled cognitive tests can only perform them on the evaluation object for the time being; the platform is not yet sufficiently developed to support these tests (for teens with social anxiety) that has been put into the account. Doctors performing evaluations are not allowed to share patient information or assessment findings. This method has proven to be incredibly successful while maintaining patient privacy.

3.1.3 Auditing user activity module

Doctors can use the qualification review tool to validate the new user's qualification information before deciding whether to approve the user's requested operation and/or proceed with the cognitive behavior evaluation function. This is completed prior to deciding whether or not to proceed with the cognitive behavior evaluation function. Similarly, on the page for reviewing qualifications, subordinate users' review status can be viewed and discussed. In this scenario, he would play the role of a doctor, performing a series of diagnostic procedures to determine the source of the problem. The attending physician will decide whether to proceed with the operation. Users of a cognitive-behavioral assessment platform designed for adolescents suffering from social anxiety may be eligible for certification. To complete their registration, newly added users must complete and submit a certification form (Garg, 2023). They will be able to use the system assessment feature once they have passed a test and been given the all-clear by a more knowledgeable physician. Even if the evaluation is unsuccessful, the new user may still be able to provide the necessary certification and check their certification history online.

3.1.4 An evaluation of cognitive behavior structure

Modern cognitive-behavioral testing infrastructure for teenage social anxiety is demonstrated. Doctors are only authorized to register, examine the credentials of their staff members, or assess patients according to their individual requirements. The steps may be accomplished by entering your login information into the necessary browser forms as specified on the relevant websites. Fig. 9.3 depicts the workflow for the cognitive behavior assessment system.

The third database design focuses on abstracting the studied data while considering the needs of the adolescents using the social anxiety and cognitive behavior assessment systems. You would not be able to access the database or any other parts of the larger infrastructure for information gathering unless you use this component of the system. The database in this system makes extensive use of relational tables. This table gives information about the users who are presently logged in (doctors). Among the information captured are the user's identity, gender, and audit status. As evidenced by

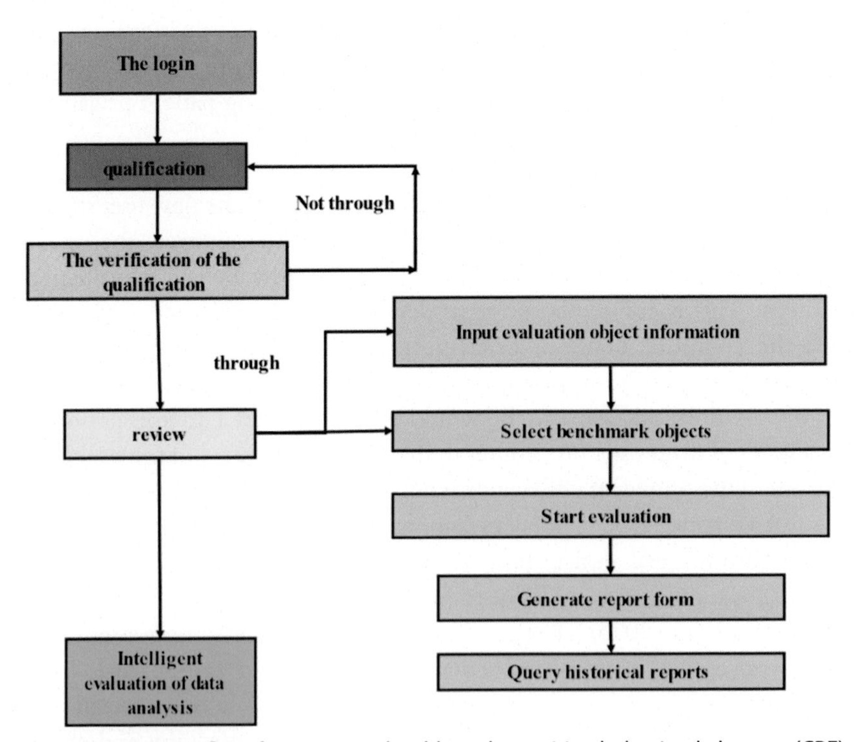

Figure 9.3 Process flow for using a cloud-based cognitive-behavioral therapy (CBE) assessment system.

Table 9.1 Physicians frequently have a core corpus of knowledge.

The data type	Primary key or not	The length of the	Null	The field
Bigint	Is	20	No	Id
Bigint	No	20	Is	The logged in user
Bigint	No	20	Is	Position
Bigint	No	20	Is	The title
Varchar	No	50	Is	The name
Tinyint	No	1	Is	Gender
Int	No	11	Is	Age
Bigint	No	20	Is	Audit doctor ID
Tinyint	No	1	Is	Review the status

the statistics in Table 9.1, physicians frequently have a core corpus of knowledge that includes.

The table shows the physicians who have treated logged-in patients are characterized in some of the most basic terms.

3.2 Historical patient data

This database stores the patient data for the user that is currently logged in. These data may contain the patient's name, address, and outpatient number. This part goes into vector learning and how to build a correlation prediction model for socially anxious teens. This method, which is based on deep learning, will help predict how CBT and social anxiety in teens work together. Many individuals, schools, families, and societal risk and protective variables for adolescent development are being studied to better anticipate and optimize the relationship between CBT. Calculating the index weight of a vector of physical traits is used to measure teenage social anxiety. As a result, the specific actions to take, as stated below, may be used to investigate the reasons for social anxiety among teenagers. It identifies the causes of adolescent social maladjustment, the social events that exacerbate it, the characteristics of the many components within each category, and the effect of membership on the attributes. Adolescents' social anxiety is often triggered by social stress, traumatic experiences, or a combination of the two. Both risk and protective variables are at work as teens develop. These are two extreme instances on the same scale. The connected variable's score for the specific adolescent decides whether a given variable is a particular factor rather than the variable's feature. Therefore, except for those pertaining to danger and protection throughout teenage growth, the importance of each variable to the algorithm is investigated. It is said that this model includes all the factors that explain why teens get social anxiety.

When psychiatric diseases are in their early, less severe stages, they might be difficult to diagnose. Furthermore, many people who are embarrassed by their illness lie about not having it. This is troublesome since many people invent their illnesses. On the other side, controlling your mental health is not difficult, and in most cases, daily stress is what causes difficulties. Because of how quickly things change in our modern society, it is more important than ever to pay serious consideration to concerns relevant to one's mental health (Kaur et al., 2022). Individually, people's behavior, tone of voice, and cardiac rate can all convey crucial information about how they are feeling. On the other side, the public has poor knowledge of the issues surrounding mental health (Ansari et al., 2021). If mental health problems are not treated and managed in a timely manner, they have the potential to develop and have disastrous consequences. Who or what must be held responsible? This endangers both individual lives and the overall quality of life in the community. An in-depth study is now being conducted to establish whether it is possible to determine whether a person is suffering from a mental illness simply by seeing their behavior or hearing their speech. Using tools powered by AI can be a powerful tool for predicting difficulties with a person's mental health. It is possible that they will look at the relationship between a person's words, behaviors, and mental health. AI-based algorithms can determine whether there is a link between specific behaviors and mental health difficulties when given adequate data. This method has the advantage of predicting when mental health illnesses will show up for the first time. It also provides the added benefit of predicting the severity of such disorders.

3.3 An overview of the CNN health intelligence Network's organization and psychological research resources

The primary purpose of this study is to determine whether specific aspects of people's behavior, voice, and pulse can predict the severity of any mental disorders they may be experiencing now. This goal will be helped by comparing participants to a group of healthy individuals. As a result, new techniques for psychological counseling and forecasting are likely to develop in the not-too-distant future. This strategy not only has the potential to forecast the onset of mental health difficulties in people based on their everyday behaviors, but it also has the potential to prevent the occurrence of those issues. Another industry that could benefit from the deployment of this technology is psychotherapy. It may be able to provide information about psychological counseling that is specifically geared to meet the requirements of various groups of people living with mental health difficulties (Skinner et al., 2019). This possibility arises since it is possible that it will be able to provide some information about psychological counseling. Many people are hesitant to seek psychological counseling from a psychologist because they value the privacy of their personal information so highly. The psychological counseling and health system developed because of this research not only anticipates future psychologically connected health difficulties that people may face, but it also delivers psychologically intelligent counseling that is personalized to the needs of everyone. When the dataset on mental and health counseling is obtained, the next phase in this project will be the digitalization of the components. The data preparation approach then employs distance correlation to discover how to divide the data into three distinct feature datasets. Each of these datasets will have a distinct set of properties. CNN and long short-term memory (LSTM) uses a range of topologies to provide psychological counseling and create health forecasts, as shown in Fig. 9.4. A substantial amount of training data are required for CNN and LSTM to understand psychological therapy and health-related components (Xiao et al., 2021). This is because each type of teaching has its own individual characteristics. To achieve this goal, the intelligent system will need access to a vast amount of information about people's mental health, which may include things like people's emotions, voices, and heart rates. Once the necessary preprocessing has been done on these data, it will be possible to feed it into the CNN. Employees at CNN will be able to better serve their viewers if they are aware of the link between the attributes and markers of psychological wellness. The general degree of physical health and the frequency of psychological difficulties are two of the most essential factors that affect a person's mental health. When the CNN's learning phase is complete, these data must be transmitted to the LSTM algorithm for

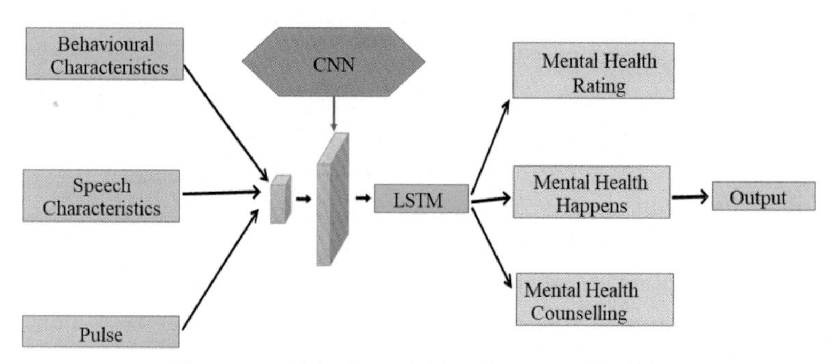

Figure 9.4 Hybrid model for the mental health.

processing. Psychiatric disorders are conditions that are linked together and worsen over time. These are conditions that worsen with time and are inextricably tied to and exacerbated by time.

These ideas can be utilized in a variety of contexts, including the overall healthcare system and mental health counseling, to name a few, as shown in Fig. 9.4 In this case, CNN and LSTM provide the solutions. The advent of mental health issues is not just due to the passage of time. While the CNN algorithm is getting better at identifying links between linguistic, behavioral, and psychological characteristics, it is still unable to extract data about time passing (Kashyap, 2019a). As a result, in addition to studying the temporal properties of speech, the authors of this study investigate how LSTM algorithms might be used to extract temporal features of people's mental health. It has been successfully implemented, and the LSTM approach is being used in several voice recognition apps. According to intelligent psychological therapy and healthcare systems, the issues people encounter with their mental health have become increasingly complex over time. At first, everyone will carry on as usual, acting as if nothing has changed. However, as time passes, the pressure imposed on it decreases. This is likely to lead to the development of psychological issues (Kashyap, 2019b). Another aspect of the emergence of mental health disorders is the fact that people respond differently to pressure. As a result, the temporal components of the mental health indicators must be retrieved using the LSTM technique in this work. The findings of the extraction of the mental health aspects will be erroneous if the LSTM technique is not used. The accuracy of the LSTM technology is responsible for these outcomes. This operation will be carried out in the paragraph that follows. This information will be delivered to the LSTM in chronological order. The data gathered can be used to study and treat mental diseases. An LSTM technique requires a structure with four gates to function properly. Once past the guards, it will carefully examine whether to spotlight key events from the past or the present (Nair, Vishwakarma, et al., 2021). As data passes through a series of gates in preparation for transmission to the next tier of the network, its relevance increases.

CNNs produce excellent outcomes when it comes to precisely mapping the various elements. Following the completion of feature extraction, the feature mapping connection can be performed. In the context of the current study, this expression refers to the relationship between people's ratings and their actions, words, and how others perceive them. This expression refers to the relationship that exists between people's ratings and overall impressions (Nair, Alhudhaif, et al., 2021). The CNN technique can complete relational mapping projects of different difficulties and has been used effectively in a variety of relational mapping use cases. If the relationships between these pieces are straightforward, adding further network levels will have little impact. The first half of this study describes the computational framework on which the LSTM technique is based. The findings of the methodology are then discussed. The fundamental goal of this system is to allow users to access crucial information at the final possible moment, right before the forgetting gate opens. It then proceeds to the next step of neural network learning after choosing which bits of the data from the previous instant to utilize based on the weight assigned to the data. This is critical if the likelihood of success is the selection criterion (Zhao & Tang, 2021). Several AI methods make use of the concept of an activation function. Only their direct association indicates the link between mental health therapy and other elements.

4. Analysis of experimental results

Experiments and simulations show that the best way to find and treat adolescent social anxiety disorder is to use deep learning to guide CBT. We chose 100 of the most socially anxious 16-year-olds to participate in our study. The predicting periods of several prediction models are compared in Fig. 9.5. We discovered that utilizing our technology to create predictions is much faster than the traditional method. Based on the results of this study, the relevance prediction model that was made may make it much easier to predict teens' social anxiety.

The accuracy of various forecast models shown in Fig. 9.5 indicates that ever since the trial, the usual prediction model has failed to predict the diagnosis or the impact of teenage social anxiety treatment with high accuracy.

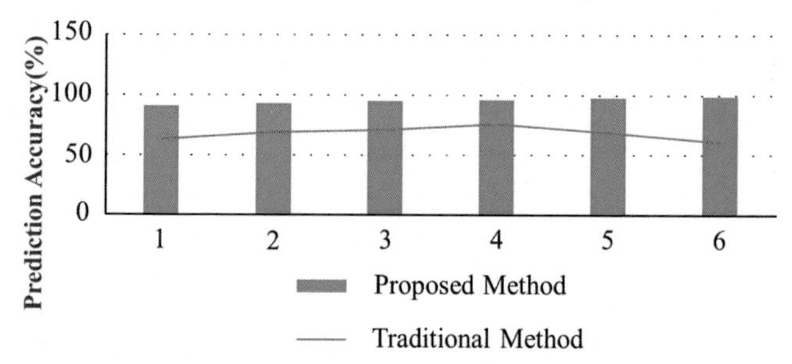

Figure 9.5 Comparative analysis of predictive accuracy of the proposed model and traditional model.

Despite the massive amount of research, general accuracy has not improved much. This study's correlation prediction algorithm has already improved forecast accuracy. Despite the rise in test volume, overall accuracy has remained consistent. The proposed model for anticipating correlations is contrasted with a more conventional method in terms of stability. The relevance prediction model utilized in the results of this investigation reveals its ability to predict the diagnosis and course of therapy for teenage social anxiety. When utilizing the strategy described in this research, the experiment's start and end points are both very consistent. Fig. 9.5 shows that the standard prediction model was 71% stable at the start of the trial when used to predict the diagnostic and treatment impact of teenage social anxiety. This proportion decreases as the number of trials increases. This work heavily employs the CNN and LSTM algorithms to investigate the properties of psychological counseling and health intelligence systems, as well as the potential uses of such a system. Changes in a person's demeanor, voice, and heart rate are routinely used to detect the onset and severity of mental health illnesses. We frequently hear comments like this. Customers can also get up-to-date information on personalized psychological therapies. The interactions between a person's speech, pulse, mental state, and psychological therapy procedures, as well as the characteristics of their psychological changes throughout time, should be tracked. People's personalities and worldviews change with time, so it is vital to keep an eye on these dynamics. The LSTM approach excels at distinguishing temporal components from mental health data. For this study, researchers separated the population of Shenzhen, China, into 2000 subgroups to assess each group's behavioral features and psychological well-being. Despite the difficulties of working and living in Shenzhen, you should have no trouble finding the information you require. Preprocessing is required as soon as the data are collected to organize the data correctly based on the relevant features. The study begins by comparing the performance of two different neural networks in feature extraction for mental well-being. It accomplishes this in the first stage by identifying the underlying association between people's mental health and their behavior using a neural network without an LSTM layer. As a result, by evaluating the person's growth across time, it is feasible to evaluate whether a given time has a discernible impact on their mental health. Fig. 9.6 depicts the disparities between actual and expected values for three components of a person's mental health. The use of an LSTM-less neural network was at the root of these problems. V1 is an example of the type of error that might occur while attempting to forecast what other people will say. The term

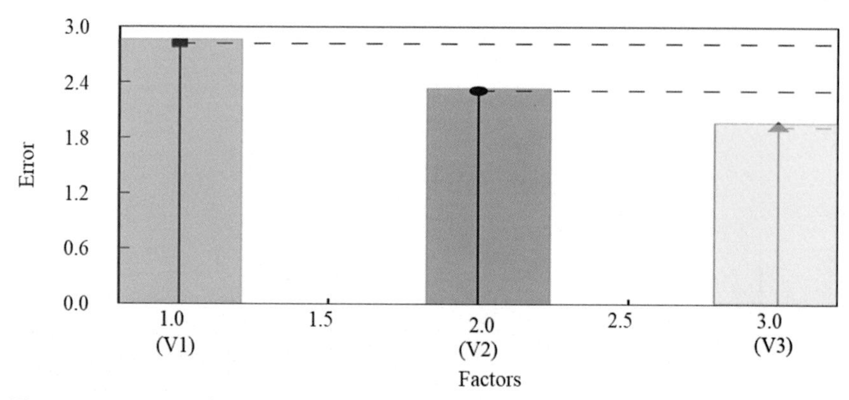

Figure 9.6 Depicts the disparities between actual and expected values for three components of a person's mental health.

"Variable V2" refers to the prediction inaccuracy that can be related to the observed behavior. When attempting to forecast the pulse-related properties of individuals, there is a certain degree of inaccuracy that might occur, and this mistake is known as V3. It has been demonstrated that the first two features have a prediction error of more than 2%. This is by far the most significant of the three errors that the approach has previously detected. Even though these differences are visible, they only account for 3% of the total. When seen through the lens of attributes connected to a person's psychological well-being, this error is good psychologically. The fact that all three of these dimensions of human nature—behavior, attitude, and character have defects makes it difficult to regard any of them as flawless. More than 2% of people intentionally mislead others with their words. A change in a person's speech patterns that is closely related to their awareness of time passing is one of the most significant indicators of how their mental health issues appear in their behavior. These features may be more common in people who have legitimate reasons to be concerned about their mental health. This could have played a role in the large inaccuracy in this area of the feature. Fig. 9.7 shows the mistakes that were made when people tried to predict three things about mental health problems.

The LSTM layer could be a good choice when trying to figure out how to extract temporal features from data related to mental health and therapy. The study also investigated the difficulties associated with adopting a mixed neural network technique. These difficulties are related to issues with mental health and therapy. This fall in accuracy is due to a reduction in the amount of noise added to the prediction process. The graph that follows this one

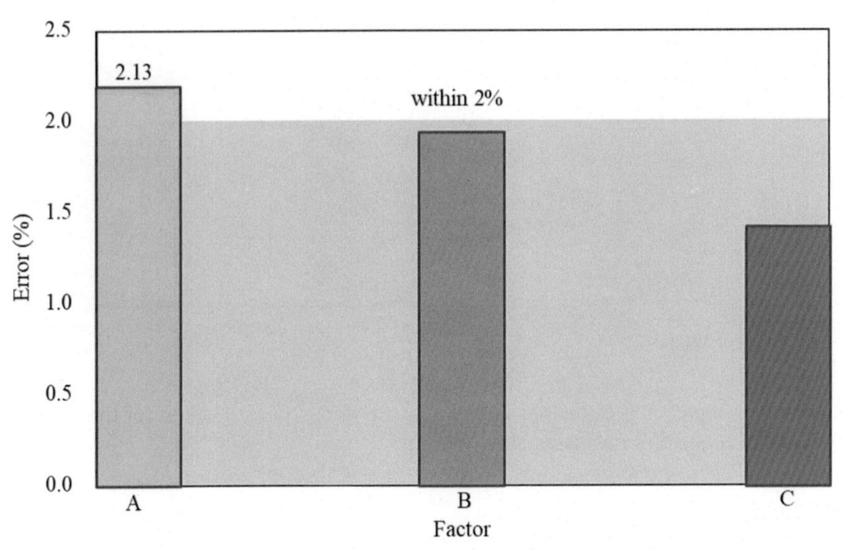

Figure 9.7 How bad long short-term memory (LSTM)-layer neural networks were at predicting the values of three attributes related to mental health.

show how far we have come since the beginning. As a result, the outlook for mental health and counseling services is substantially improving. This demonstrates beyond a doubt that concerns regarding mental health and therapy issues have clear time components. Counseling is a tried-and-true method for promptly addressing concerns, and its efficacy has been shown beyond a shadow of a doubt. This highlights how the effects of time on therapy and mental health are inextricably linked and cannot be considered independent of one another. Counseling sessions for mental health and drug addiction concerns are usually longer than those for other types of diseases. The prediction error for speech features, which have more temporal components than other mental health indicators, fell from 2.8% to 2.13% during the trial. This was the case because speech features included more temporal components than other mental health indicators. The amount of incorrect data on behavioral variables connected with mental health has decreased from 2.29% to 1.94% in the last year. After adopting some of the suggested changes, the accuracy of the pulse feature's forecasts decreased from 1.87% to 1.41%. It is regarded as a highly important component of the offer, like the preceding aspect of the discount, which means that it is a significant portion of the overall package. This demonstrates unequivocally that the LSTM approach is capable of effectively predicting a wide range of psychological well-being traits shown in Fig. 9.7 This is caused by the fact that it can now predict human behavior and pulse characteristics with an error rate of less than 2%. When it comes to early diagnosis of mental health difficulties and successful coupling with psychological counseling, having a prediction error distribution that contains these three traits is quite advantageous. Furthermore, this type of distribution helps reduce the likelihood of false positives. As a result, it is critical to properly address time constraints while attempting to foresee parts of mental health and therapy.

Given this backdrop, it should be evident that individuals' speech characteristics influence their mental health and the effectiveness of counseling intelligence systems. The discussion of linguistic differences among diverse groups of people makes it easy to conclude that this is the case. In this work, we compared the accuracy of 20 distinct sets of speech features' prediction mistakes. It is reasonable to expect that estimates about the efficacy of psychotherapy and normal speaking will follow the pattern depicted in Fig. 9.7. The green region depicts the prediction error for mental and physical counseling that is within 2% of the general average. This makes it easier to understand how the prediction error distribution is distributed among the various linguistic features. Most speech feature errors are in the 2% range and frequently occur in groups. Only three datasets had prediction errors of more than 2%. This is a negligible proportion of the total number of test items distributed (indicated in red). The fact that a small percentage of the error distribution falls inside the 1% range is the evidence that CNN and LSTM algorithms can understand the language components of mental health and therapy more precisely. To compile these data, the linguistic characteristics of people's speech were investigated. Trying to predict a person's mental health based on this aspect of their personality is one of the most difficult things to do. When attempting to predict a person's behavioral and pulse traits, the margin of error was less than 2%. This level of accuracy was amazing. This error distribution can aid in the early detection of mental health issues and, if necessary, the fast and competent psychological treatment of patients suffering from such conditions. The study approaches the subject of mental health and therapy in a trustworthy and insightful manner, one that is not only highly applicable but also generally applicable. People can be educated to the point where they can predict mental health issues, and this knowledge may also aid in preventing psychological disorders. When it comes to accurately forecasting the characteristics of psychological and health counseling, the CNN-LSTM method outperforms other methods. This has unintended consequences for the likely routes that mental health practices and counseling will take in the next few years. This study is interesting because it includes approaches from the field of artificial intelligence and predicts traits vital in psychological and health therapy. It is expected to be useful in assisting people in both the prevention and treatment of mental health disorders.

5. Conclusion

Predicting how well CBT will work can help treat this growing mental health problem. This treatment is currently the gold standard for treating this problem. This is a necessary step that must be taken to adequately handle mental health conditions. This is only one of several deep-learning applications that are now being employed in society. The first method is to raise teenagers' social anxiety. Second, using a multiobjective evolutionary algorithm, a model is built to predict how well CBT will work for teenagers with social anxiety. According to the data presented here, anxiety disorders have a devious clinical course, a poor possibility of recovery, and a high recurrence rate. When specific comorbid mental difficulties persisted, the chance of recovery was lowered and the risk of repeated anxiety disorders was raised. Several personal, interpersonal, institutional, and societal perspectives are used to explore the impact of risk and protective variables on adolescent development. The gray correlation coefficient may be used to differentiate the subjective and objective features of adolescent social anxiety. A model was also developed to estimate how effectively cognitive behavioral therapy would reduce teens' social anxiety. The strategy given in this research is more valuable and successful than competing alternatives for supporting young people in the development of their social skills. These results are useful because they expand our understanding of how to diagnose and treat these illnesses. This study makes extensive use of two well-known neural network-working methodologies, CNN and LSTM. It can complete the feature mapping procedure and extract psychotherapy and mental health parameters. The purpose of this study is to compile a history of mental health evolution over many generations by examining the correlation between people's speech patterns, heart rates, and other behaviors. These three qualities are crucial markers that can help determine whether a person has a mental health problem. Another issue of interest examined in this study was how the presence or absence of an LSTM neural network affects one's ability to predict attributes related to mental health and therapy. The goal of this study was to answer the following question: There were times when an artificial neural network was employed throughout the investigation of this issue and other times when it was not. The precision of forecasts provided by neural networks using LSTM layers is one facet of the network's overall performance. Future prognostications will center on issues concerning a person's mental health. Individual behavioral features related to mental health were identified as the most important identifiers of subgroups. When it comes to forecasting outcomes in the disciplines of mental health and counseling, both the CNN method and the LSTM approach have an error rate of no more than 2.13% when using their respective models.

References

Al-Abyadh, M. (2010). The effectiveness of a meaning therapy program in reducing the degree of social phobia and improving the meaning of life for a sample of young people. *The Scientific Journal of the College of Education, 10* (Original work published 2010).

ALHarbi, A. A., Al Ahmadi, Ahmed F., Al Saedi, Mohammed F., Alghamdi, Khalid, Radman Al-Dubai, Sami A., & Muhalhil, Rawan E. (2022). Prevalence of depression, anxiety and stress and associated factors among nurses working in a tertiary hospital in Al-Madinah city 2021, a cross sectional study. *International Journal of Advanced Research, 10*(05), 837—845. https://doi.org/10.21474/ijar01/14780

Ansari, G., Garg, M., & Saxena, C. (2021). *Data augmentation for mental health classification on social media.* arXiv. https://arxiv.org.

Cao, Xingping, Luo, Zeyuan, Qiu, Junlin, & Liu, Yan (2022). Does ostracism impede Chinese tourist self-disclosure on WeChat? The perspective of social anxiety and self-construal. *Journal of Hospitality and Tourism Management, 50,* 178—187. https://doi.org/10.1016/j.jhtm.2022.02.013

Fitzgerald, Amanda, Rawdon, Caroline, O'Rourke, Claire, & Dooley, Barbara (2019). Factor structure of the social phobia and anxiety inventory for children in an Irish adolescent population. *European Journal of Psychological Assessment, 35*(3), 346—351. https://doi.org/10.1027/1015-5759/a000420

Garg, Muskan (2023). Mental health analysis in social media posts: A survey. *Archives of Computational Methods in Engineering, 30*(3), 1819—1842. https://doi.org/10.1007/s11831-022-09863-z

Gómez, Carmen, Redolat, Rosa, & Carrasco, Carmen (2017). Bupropion induces social anxiety in adolescent mice: Influence of housing conditions. *Pharmacological Reports, 69*(4), 806—812. https://doi.org/10.1016/j.pharep.2017.03.010

Guo, Yuanyuan, Lu, Zhenzhen, Kuang, Haibo, & Wang, Chaoyou (2020). Information avoidance behavior on social network sites: Information irrelevance, overload, and the moderating role of time pressure. *International Journal of Information Management, 52.* https://doi.org/10.1016/j.ijinfomgt.2020.102067

Kashyap, R. (2019a). Big data analytics challenges and solutions. *Big Data Analytics for Intelligent Healthcare Management*, 19—41. https://doi.org/10.1016/b978-0-12-818146-1.00002-7

Kashyap, R. (2019b). Security, reliability, and performance assessment for healthcare biometrics. *Advances in Medical Technologies and Clinical Practice*, 29—54. https://doi.org/10.4018/978-1-5225-7525-2.ch002

Kaur, S., Bhardwaj, R., Jain, A., Garg, M., & Saxena, C. (2022). Causal categorization of mental health posts using transformers. In *ACM international conference proceeding series* (pp. 43—46). https://doi.org/10.1145/3574318.3574334. Association for Computing Machinery.

Keil, V., Tuschen-Caffier, B., & Schmitz, J. (2022). Effects of cognitive reappraisal on subjective and neural reactivity to angry faces in children with social anxiety disorder, clinical controls with mixed anxiety disorders and healthy children. *Child Psychiatry and Human Development, 53*(5), 886–898. https://doi.org/10.1007/s10578-021-01173-y

Laczkovics, C., Kothgassner, O. D., Felnhofer, A., & Klier, C. M. (2021). Cannabidiol treatment in an adolescent with multiple substance abuse, social anxiety and depression. *Neuropsychiatrie, 35*(1), 31–34. https://doi.org/10.1007/s40211-020-00334-0

Mannarini, Stefania, Balottin, Laura, Toldo, Irene, & Gatta, Michela (2016). Alexithymia and psychosocial problems among Italian preadolescents. A latent class analysis approach. *Scandinavian Journal of Psychology, 57*(5), 473–481. https://doi.org/10.1111/sjop.12300

Mills, J. A., & Strawn, J. R. (2020). Antidepressant tolerability in pediatric anxiety and obsessive-compulsive disorders: A bayesian hierarchical modeling meta-analysis. *Journal of the American Academy of Child & Adolescent Psychiatry, 59*(11), 1240–1251. https://doi.org/10.1016/j.jaac.2019.10.013

Moser, J. S., Huppert, J. D., Duval, E., & Simons, R. F. (2008). Face processing biases in social anxiety: An electrophysiological study. *Biological Psychology, 78*(1), 93–103. https://doi.org/10.1016/j.biopsycho.2008.01.005

Nair, R., Alhudhaif, A., Koundal, D., Doewes, R. I., & Sharma, P. (2021). Deep learning-based covid-19 detection system using pulmonary CT scans. *Turkish Journal of Electrical Engineering and Computer Sciences, 29*(SI-1), 2716–2727. https://doi.org/10.3906/elk-2105-243

Nair, R., Vishwakarma, S., Soni, M., Patel, T., & Joshi, S. (2021). Detection of covid-19 cases through X-ray images using hybrid deep neural network. *World Journal of Engineering, 19*(1), 33–39. https://doi.org/10.1108/wje-10-2020-0529

RAN, Guangming, LI, Rui, & ZHANG, Qi (2022). Neural mechanism underlying recognition of dynamic emotional faces in social anxiety. *Advances in Psychological Science, 28*(12), 1979–1988. https://doi.org/10.3724/sp.j.1042.2020.01979

Ran, Guangming, Li, Rui, & Zhang, Qi (2022). The effect of attention bias modification on the recognition of dynamic-angry faces in individuals with high social anxiety: Evidence from event-related brain potentials. *Current Psychology*, 1–12. https://doi.org/10.1007/s12144-022-03303-8

Şimşek, Caglar, Yalcınkaya, E. Y., Ardıc, E., & Yıldırım, E. A. (2020). The effect of psychodrama on the empathy and social anxiety level in adolescents. *Turkish Journal of Child and Adolescent Mental Health, 27*(2), 96–101. https://doi.org/10.4274/tjcamh.galenos.2020.69885

Skinner, D., Kendall, H., Skinner, H. M., & Campbell, C. (2019). Mental health simulation: Effects on students' anxiety and examination scores. *Clinical Simulation in Nursing, 35*, 33–37. https://doi.org/10.1016/j.ecns.2019.06.002

Song, Xiaokang, Song, Shijie, Zhao, Yuxiang (Chris), Min, Hua, & Zhu, Qinghua (2021). Fear of missing out (FOMO) toward ICT use during public health emergencies. *Journal of Database Management, 32*(2), 20–35. https://doi.org/10.4018/jdm.2021040102

Swee, Michaela B., Taylor, Wilmer M., & Heimberg, Richard G. (2021). *Cognitive behavioral therapy for social anxiety disorder.* Cambridge University Press (CUP). https://doi.org/10.1017/9781108355605.007

Vincent, A., Heima, M., & Farkas, K. J. (2020). Therapy dog support in pediatric dentistry: A social welfare intervention for reducing anticipatory anxiety and situational fear in children. *Child and Adolescent Social Work Journal, 37*(6), 615–629. https://doi.org/10.1007/s10560-020-00701-4

Waldron, S. M., Maddern, L., & Wynn, A. (2018). Cognitive-behavioural outreach for an adolescent experiencing social anxiety, panic and agoraphobia: A single-case experimental design. *Journal of Child and Adolescent Psychiatric Nursing, 31*(4), 120–126. https://doi.org/10.1111/jcap.12216

Xiao, W., & AF, A. F. (2021). *Why individuals with alexithymia symptoms more likely to mobile phone addicted? The multiple mediating roles of social interaction anxiousness and boredom proneness*. https://doi.org/10.26226/morressier.618aaeaa4a84e7b4701d81db. Original work published 2021.

Xiao, R., Zhang, C., Lai, Q., Hou, Y., & Zhang, X. (2021). Applicability of the dual-factor model of mental health in the mental health screening of Chinese College Students. *Frontiers in Psychology, 11*. https://doi.org/10.3389/fpsyg.2020.549036

Zhao, Y., & Tang, Q. (2021). Analysis of influencing factors of social mental health based on Big Data. *Mobile Information Systems, 2021*, 1–8. https://doi.org/10.1155/2021/9969399

CHAPTER TEN

Human AI: Ethics and broader impact for mental healthcare

Suyesha Singh[1], Ruchi Joshi[1], Paridhi Jain[1] and K. Abhilash[2]
[1]Department of Psychology, Manipal University Jaipur, Dahmi Kalan, Rajasthan, India
[2]Amity University Rajasthan, Kant Kalwar, Rajasthan, India

1. Introduction

The 21st century has witnessed numerous technological, medical, educational, and social advances. Dedicated research has proved extremely beneficial in providing treatment and management of various health issues; physical and mental. However, while there is a sense of readiness for seeking the latest treatment among patients and sympathy for people suffering from physical conditions and disorders, stigma, and prejudices are associated with mental health conditions. While physical conditions are visible and usually family, friends, and society are sympathetic and forthcoming with support, mental health problems often appear invisible, and there is a sense of reluctance for admitting the same both by the persons directly affected and people around them. Unfortunately, the stigma is both social and self-stigma.

Mental health diseases are the leading cause of individuals experiencing feelings of inadequacy, and it accounts for spending almost 32% of years with a disability worldwide (Vigo et al., 2016). Furthermore, there is a rapid spike in mental illnesses after the pandemic; substance abuse, suicide rates, depression, and anxiety cases increased in the last few years (Jeste et al., 2020). According to Blease et al. (2020), stigmatization of psychiatric issues, a lack of funds and resources, and a shortage of qualified mental health care professionals are the main obstacles to treating mental health problems. In the coming decades, it is anticipated that cultural and demographic shifts would strain the world's already finite supply of services concerning one's physical and mental wellness. According to the WHO, 144 low- and middle-income nations are expected to lack 1,180,000 mental health care professionals, comprising 55,000 psychiatrists, 628,000 psychiatric nurses, and 493,000 social workers necessary to address psychiatric diseases (Scheffler et al., 2011). It will result in a rise in difficulties like excess urbanization,

Emotional AI and Human-AI Interactions in Social Networking
ISBN: 978-0-443-19096-4
https://doi.org/10.1016/B978-0-443-19096-4.00005-5

population aging, pollution, migration, and increased substance use, which will lead to mental health problems. In such situations, artificial intelligence can prove extremely useful in providing support to people with psychiatric issues.

The inclusion of artificial intelligence in the healthcare industry is growing, although there are still only a few applications for it within the domain of mental health services. There is a great deal of opportunity in the realm of mental health treatment using AI. To comprehend the psychopathology of psychiatric issues accurately remains a challenging task for mental health practitioners, and the use of AI for the same can prove very beneficial for both the parties concerned, the practitioners, and the patients and clients. Researchers can quickly and efficiently determine the prodromal stage of a disease, identify risk factors, create, and choose an appropriate treatment by employing AI-based technologies. Designing a unique preventative strategy requires early disease detection. Considering current situations, mental health apps, software, and chatbots can make mental health services more accessible and comfortable in seeking help which is often avoided by affected people due to the stigma attached to getting consultation from mental health service providers. It also proves as an effective means to provide psychoeducation, providing immediate emotional support in case of crisis due to anonymity and fear of being judged by the therapist/psychologist (Vilaza & McCashin, 2021). According to Mahajan (2020), the healthcare industry of AI will grow from five billion dollars in 2020 to 45 billion dollars by 2026. There will also be an increase in the implementation of Electronic Health Records (EHR) in several countries (Evans, 2016). Currently, the use of AI in mental healthcare services is limited. To better instruct the therapeutic applications, the computing capacity tapped by AI systems could be used to disclose the intricate pathophysiology of mental health conditions (Graham et al., 2019).

2. Status of mental health issues

According to WHO, one in six persons experience some form of mental illness, which places a financial, emotional, and social strain on both the patient and the caregivers. Around 970 million individuals or nearly 10.7% of the global population, experience some form of mental illness. Females are more affected by mental illness and mental health issues (11.9%) than men (9.3%) worldwide (Dattani et al., 2018). The prevalence of anxiety disorder is almost 4% (284 million), 3.5% (264 million) depression, 1.38%

(107 million) alcohol abuse, 0.74% (71 million) substance abuse disorder, 0.51% (46 million) bipolar disorder, and 0.3% (20 million) schizophrenia worldwide (Dattani et al., 2018).

As mentioned above, there is a serious paucity of trained mental healthcare professionals worldwide, and the gap between service providers and those seeking assistance is widening, especially in developing countries due to a lack of resources, economic strains, and stigma connected to mental health problems.

According to the National Mental Health (NMH) survey estimation, there were 10.6 mental health concerns per 100 people in India in 2016. Approximately 150 million individuals are suffering from mental health issues and need immediate treatment (Hariharan et al., 2020). Almost 37% which is one-third of the global suicide among females and 24% of global suicide among males occurs in India (Vos et al., 2017). According to (Srivastava et al., 2016), India had a suicide rate of 16 per 100,000 people in the year 2015. In India, there are less than 1% of mental healthcare professionals available for the whole country (Garg et al., 2019). Compared to the WHO standard of one psychiatrist for every 100,000 people, there are just 0.3 psychiatrists for 100,000 people. For every 100,000 individuals, there are 0.07 social workers, 0.12 nurses, and 0.07 psychologists available (World Health Organization, 2018). To bridge this gap, AI can be used for the intervention and treatment of psychiatric illness, especially in India. Owing to the stigmatization, lack of awareness, and severe paucity of trained professionals the world over, automation of apps/software related to mental health is the urgent need of the hour.

3. AI in daily affairs

We have entered a new era which is known as the "Digital revolution" (Pang et al., 2018). Digital technology is an inevitable part of human lives. Digital technology has changed the communication system, information-seeking system, and almost all areas of individual functioning. No aspect of life is not touched by artificial intelligence, right from assessing and sharing information, purchasing, social interactions, research, academics, and last but not the least, rather most importantly healthcare. One major example of the digital revolution is artificial intelligence, first used in 1956 (Schwab, 2017). John McCarthy is credited with coining the phrase "artificial intelligence," a computer scientist in the 1950s. According to him "the science and technology of creating intelligent machines are known as AI"

(McCarthy, 1989). Alan Turing also authored a paper at the same time titled "Computer Machinery and Intelligence" due to his work he is another father of artificial intelligence (Turing, 2007). He suggested that in the future, machines will become intelligent. Currently, AI is present in the everyday life of the masses. The healthcare industry has also welcomed and adopted the use of AI for rendering their services (Hengstler et al., 2016). The application of AI in the field of mental health treatment is currently in the preliminary phase.

4. Application of AI in mental health services

AI at present is employed in improved illness diagnostics and timely detection of diseases, advancement, cure, and prevention. The physical healthcare sector started using AI-based services readily and more frequently in comparison to mental healthcare services (Jiang et al., 2017; Miller & Brown, 2018). One crucial aspect of AI is the ability to rapidly analyze huge data sets where algorithms can sometimes outperform skilled physicians in terms of performance. Despite the accuracy and speedy analysis of the identification, treatment, and prevention of diseases, it is very unlikely that AI would overtake human presence. There is a reluctance in the involvement of AI in healthcare unless accompanied by rigorous research and extremely careful vigilant human presence. While the use of AI in physical health applications is rapidly increasing, the mental health discipline is slow to adapt.

Mental healthcare professionals work more extensively using the client-centric approach in their practices, they depend more on soft skills such as building rapport with the client and being more empathetic with the client (Gabbard & Crisp-Han, 2016). The data in the mental healthcare services are more qualitative and subjective, but AI can still be beneficial in the mental healthcare sector (Janssen et al., 2018; Luxton, 2014). AI in the mental healthcare industry can help in understanding various psychiatric illnesses in-depth and redefining diagnosis (Bzdok & Meyer-Lindenberg, 2018). With the help of AI, clinicians can easily understand the biopsychosocial underpinnings of every client (Jeste et al., 2019). Although, using AI in understanding biopsychosocial systems is difficult due to heterogeneity in the pathology of the development of psychiatric illness, using AI in the mental healthcare industry can help in developing better prediagnostic assessment and screening tools, formulating a predisposition model, and planning effective treatment (Shatte et al., 2019). Following are some applications of AI in the field of mental health care.

4.1 Machine learning

It is the most commonly employed branch of artificial intelligence in the healthcare sector. It helps in estimating or forecasting results from new data or probable future occurrences that can be derived through data using data-driven algorithms (Shatte et al., 2019). There are suggestions for applying ML to data derived from the mental health care sector. The International Society for Bipolar Disorders Big Data Task Force recommends employing ML for different avenues for identification, treatment, and management of the disorder; right from collecting information of clients through various genetic bases to working on the production of risk calculators to help people suffering from BPD (Passos et al., 2019). Similarly, Park et al. (2019, pp. P342—P343) focused on different empirical research strategies for assessing the effectiveness of AI-based models in healthcare. AI is ideally equipped to manage precise, digitized, or "big data" (Bzdok & Yeo, 2017). In the discipline of psychiatry, deep learning (DL), a branch of machine learning, offers a lot of potentials (Durstewitz et al., 2019). Utilizing the largest publicly accessible datasets deep learning has been used to facilitate diagnosis, notably for dementia (Vieira et al., 2017). According to Garg (2023), using social media posts as a basis, machine learning, and deep learning techniques can be used to detect suicidal tendencies. As a result, it will be simpler to recognize anxiety, stress, depression, and other clinical syndromes based on a person's use of social media. EHRs, smartphone electronic phenotyping data, and social media networks are record-high data sources that potentially supply the inputs required to properly utilize deep learning for mental health apps.

4.2 Natural language processing

NLP consists of ways of changing texts from unstructured to structured formats for further aiding and assisting the studies related to recognition of voice, different emotions, vocabulary, optical characters, etc., for mental health promotion and diagnosis of various disorders (Sheikhalishahi et al., 2019). NLP is increasingly being used in the medical field, especially in psychiatry. Graham (2019), Koleck et al. (2019) advocated the use of NLP in the application of psychiatry by enlisting its benefits such as cost-effectiveness, easy accessibility, and ability to investigate minute details that enable it to focus on phenotypes as well as comorbidities. In one of the few instances that prove its effectiveness the application of NLP in the Clinical Record Interactive Search (CRIS) Platform helped in predicting suicidal ideation as well as the possibility of relapse of psychiatric symptoms, demonstrating its value and utility in prediction, diagnosis, management, and treatment of mental health problems (Kaur et al., 2022).

4.3 Helpful in diagnosis

Multiple crucial techniques exist for AI systems these days to assist with differentiated diagnostic issues. As an example, unipolar versus bipolar can be distinguished using features of neuroimaging studies and brain scans (Redlich et al., 2014), and different types of dementia could be distinguished using structural MRI scans (Dwyer et al., 2018). The ability to discriminate among diseases with identical early diagnostic manifestations with alternative therapeutic modalities can be improved in part by artificial intelligence techniques (Kloppel et al., 2008). The diagnosis of new ailment subgroups may also be aided by information derived from computer algorithms based on a wide range of clinical presentations, demographic traits, and environmental factors (Dwyer et al., 2018). Thirdly, AI approaches may generate models from novel or unusual data sources by assembling information from many disparate data streams and sources.

4.4 Helpful in prognosis

Accurate prognoses for psychiatric patients can be strengthened by AI approaches used with longitudinal studies (Dwyer et al., 2018). Studies have simulated the course of depression using genomic, linguistic, neuroscience, and EHR data, suicidal behavior, future substance misuse, and functional outcomes (Schmaal et al., 2015; Tran et al., 2014). Without largely relying on psychiatric theories currently in use, AI algorithms can also use information strategies to create novel clinical risk predictive models (Fusar-Poli et al., 2018).

4.5 Helpful in deciding treatment

AI applications have a lot of potentials to support clinical psychiatric therapies in various ways. First, AI can be utilized to anticipate treatment outcomes, thereby avoiding moment psychotherapies, invasive neurostimulation treatments, and failed drug trials (Fusar-Poli et al., 2018). Second, AI techniques can aid in predicting severe therapeutic adverse effects (Passos & Mwangi, 2018). In one research, lithium-treated participants' EHR data were utilized to forecast the development of renal insufficiency (Castro et al., 2015). Thirdly, novel conceptualizations of pathogenic mechanisms can be developed with the use of AI methods. According to one study, bipolar illness is a neuroprogressive condition that exhibits anomalies in neuroimaging, and differences have been observed in bipolar clients in comparison to the control group (Mwangi et al., 2016).

Clients with hallucinations converse with an avatar of their alleged oppressor, which makes the avatar less antagonistic and makes therapy more convenient both for the client and the therapist. The avatar is voiced by the therapist. In a study published in 2018, Craig et al. (2018) examined how AVATAR therapy affected auditory verbal hallucinations and found it to be more effective than supportive counseling. Garg (2023) explains the importance of new frontiers in terms of promoting and maintaining mental health with the help of factors such as the happiness index and ease of living index of the native's country, the differences due to geographical locations, and behavioral analysis for instance usually the posts related to depressive tendencies are mostly expressed late at night than in mornings and approach based on linguistics.

There are various new treatment modalities in AI-based approaches, but still, there is a lack of ethical and governing bodies to take care of these services.

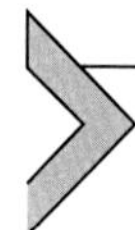

5. Issues of employing AI ethically in the mental health care services

5.1 Regulation and governance of AI based apps and software

Mobile applications based on AI are not subject to any regulations. According to research from 2015, the United States had more than 45,000 AI-based mental health applications. These applications' trustworthiness and validity are in doubt because the majority of them have not been evaluated, and the few that have been validated mainly did so with a small group and for a brief period. Due to the poor quality of the content and unsafe guidance provided, this type of service poses a risk to users who use these apps for mental health treatments.

These apps mostly offer self-management tips and lifestyle recommendations, many of which are not regarded as "medically right" (Shen et al., 2019). It is crucial to have a specific regulatory body in place to oversee these services. The "National Health Service Digital Health Library," which includes a list of reliable and secure apps, was produced in the United Kingdom by National Health Services.

Martinez and Dasgupta (2020) stressed the need of working on privacy and data protection aspects of the digitalization of mental health apps and services during the pandemic. While mental health applications and software delivered promising results in providing assistance and remedies during a pandemic and the leniency in regulations for the same had also proved beneficial for clients/patients and mental health professionals in short term, it is extremely necessary that rights and well-being of clients, therapists, and all stakeholders concerned are taken care of within ethical and legal framework that is unanimously accepted.

5.2 Uncovering AI literacy

To use AI in a righteous way, it is important that algorithms used in detecting psychiatric illnesses must be accurate and must not increase any risk for clients. The gap in AI literacy must cover the literacy gap between clients and clinicians. When using the information provided by someone else, data scientists might not have the essential knowledge of the data's complexity to recognize potential limitations. However, there are not set standards for the usage of artificial intelligence in mental healthcare services (Nebeker et al., 2017).

5.3 The issue of privacy

Services for mental healthcare using artificial intelligence are freely available, but people need to check these service providers' privacy policies, app functionality, and discrimination policies. This might assist in avoiding choosing a subpar or improper app that might compromise the client's privacy, confidentiality, or general care. While automation of mental health services can prove beneficial, it should however be ensured that issues like privacy, personal data being traded, and sensitivity toward cultural, racial, and societal norms should be ensured (Vilaza & McCashin, 2021). In their study, epistemic, normative, and overarching ethical challenges were identified by Morely and associates (Morley & Floridi, 2020). Rubeis (2022) explained three important elements of ethical concerns about the use of AI for mental healthcare, namely, data mining, self-monitoring, and Ecological Momentary Assessment (EMA).

5.4 Consent in record keeping

The record of mental health services is delicate since stigma, prejudice, and fear are linked to it. Before implementing a new, supposedly effective treatment method, the mental healthcare provider must earn the client's trust. Misuse of data occurred in some instances. For instance, if a client loses their capability owing to some mental health condition, it is unclear if their preliminary consent to passive surveillance will remain authentic. When using AI in the treatment, the client would also need to give consent for using a higher amount of data, some clients give consent for medical notes but not for another type of monitoring such as video, audio, and another type of record keeping.

5.5 Lack of human interaction and feedback

Numerous AI apps are made to give users information without being directly accessible by a person. While the user may benefit from this, there are also drawbacks. Humans are sociable creatures who frequently gain from interacting with others. Utilizing AI apps exclusively can decrease the chances of social contact. This may be an issue, particularly for those who lack social abilities or are isolated and lonely. In contrast, certain AI applications are made to give feedback automatically. Although some individuals might find this useful, others might find it harmful. Human feedback is frequently valuable because it enables users to learn from their errors and alter their behavior as they go. Without human input, a user would not be aware of their errors and may not have the chance to do better next time.

5.6 Lack of control over AI

An expert authority does not control applications either. Because of this, people who use them for mental health services frequently do not receive the same level of care that would be offered by a qualified mental health professional. Instead, consumers might be utilizing a program built by programmers that have not undergone the same level of scrutiny as a mental health expert would. People do not have the same legal safeguards when using mental health apps as they do when they visit a qualified mental health expert.

5.7 Training for mental healthcare professionals

Mental healthcare workers must have sufficient training before using any AI-based services for mental health because it can help in reducing human biases.

5.8 Importance of face-to-face treatment

Individuals cannot completely rely on AI-based technology. These are helpful tools, but these cannot completely replace face-to-face treatment services, in-patient services, and other important aspects of mental health treatment as it requires an idiosyncratic approach to the information collection, prognosis, diagnosis, and treatment plan. The genetic, socioeconomic, cultural, and personality profile of each individual is unique and overgeneralization for treatment can lead to disastrous consequences.

6. Challenges in using AI in the mental healthcare setting

It is challenging to create AI-based mental healthcare services that are successful as several potential biases in the treatment of mental illness do come up frequently. These potential biases are.

1. Overreporting and underreporting by the patient
2. Preferring face-to-face treatment modality
3. Privacy, data breach, and confidentiality related issues
4. Lack of research evidence of AI based mental health services
5. Lack of psycho-diagnostic assessment and general physical examination
6. Higher chances of misdiagnosis
7. Lack of understanding of cultural, social, and familial issues
8. Lack of representative data
9. Lack of reliability and validity
10. Difficulty in managing high-risk situations
11. Algorithm biases
12. Stigma related to help-seeking behavior.

7. Legal and ethical considerations in AI in mental healthcare services

While the role of psychotherapy is to ensure the overall well-being of the client seeking consultation, certain ethical aspects are taken into consideration while entering a professional relationship. While administering psychotherapy the principles of autonomy, beneficence, nonmaleficence, and justice are guiding principles for therapists. Carl Rogers (1957) emphasized unconditional regard and respect for the client. Maslow (1943) gave the principle of self-actualization stressing the need for therapists to be nonjudgmental and the ability of the client to decide for themselves the best option for the solution to their problems and capacity to develop to the maximum. Keeping the above principles in mind while automation of mental health services can prove beneficial, it should however be ensured that issues like privacy, personal data being traded, and sensitivity toward cultural, racial, and societal norms should be ensured (Vilaza & McCashin, 2021).

There is a lack of country-wise legal, and ethical rules and guidelines for use of AI in mental healthcare services, in this situation, the application of the FATE model serves as a guiding light for use of AI in mental health issues. By following the FATE model, AI can be used in righteous ways in various sectors including mental health services. FATE refers to "Fairness, Accountability, Transparency and Explainability." The summary of the FATE model is explained in Table 10.1.

Apart from the application of FATE analysis, the American Psychological Association elucidates the principles of counseling that need to be kept in mind while applying AI for mental health issues. "The American Psychological Association" has laid five fundamental values for psychologists, including benevolence and nonmaleficence, integrity and accountability, honesty, fairness, and respect for others' rights, freedoms, and dignity. Fulmer et al. (2021) describe six ethical principles that are encountered by therapists or agencies while engaging AI in mental healthcare, as follows: boundaries of competence, limited ethical codes, transparency, cultural diversity, reliability and validity, and cyber security. The therapist or practitioner must be competent in not just the above-mentioned principles but also have enough training and exposure to the latest technological advancements.

Similarly, the Lord's report on AI principles mentioned that artificial intelligence should be used for the betterment of humanity; it should be fair and intelligent, should not be made powerful to harm others, dismantle, or purposely mislead humans, should grant all citizens the right to education so they can flourish mentally, emotionally, and economically alongside artificial intelligence, and should not be used to replace human intelligence.

"Canada Protocol" is a checklist that guides AI for dealing with and regulating mental health issues and aids in suicide prevention, and can serve as guidelines for practitioners and policymakers. The protocol emphasizes biases, transparency, the privacy of people, safety, and security, mental well-being risks, and the specification of the autonomous intelligent system (Mörch et al., 2019).

Table 10.1 FATE model for Legal and ethical consideration in AI in the mental health care services.

S.No.	FATE model	Summary	Instances in AI-based mental health services
1.	Fairness	AI services need to be unbiased and discrimination-free. Fairness in AI services must include lawfulness, the privacy of data, protection of data, and removal of algorithm biases.	Fairness in AI-based mental health services can be applied when these services are algorithm-biased free services, gender disparities free, data is encrypted, and bias-free on a culturally diverse population.
2.	Accountability	Accountability in AI refers to taking responsibility for AI errors. These errors can be minimized in 2 ways: first is, human-in command system, in which human authority takes control of AI-based services. Second is organizational accountability in which companies take responsibility for any AI-based errors.	In certain cases, AI-based services can misdiagnose an individual. In those situations, which can create problems for the client. In that kind of situation, any accountable authority such as a psychiatrist, or psychologist needs to give the right direction to the user.
3.	Transparency	Transparency in AI refers to explaining itself, its working process, predictive nature. Earlier AI-based services are considered black box services due to lack of transparency in the system but currently, the adequacy of training data makes it more transparent.	Before using any AI-based services, users can get access to all the relevant terms and conditions, potential merits, and demerits. Transparency in AI-based mental health services can be done when the data is secure, privacy is maintained, and policies are clearly stated.

Table 10.1 FATE model for Legal and ethical consideration in AI in the mental health care services.—cont'd

S.No.	FATE model	Summary	Instances in AI-based mental health services
4.	Explainability	It refers to the concept that AI-based technology can be easily understood by humans. AI services will become more acceptable when it is more explainable to the person who is using them.	All the AI-based services need to be easily understandable so that users can easily use these services. Additionally, AI-mental health services become more explainable when it is clear, parsimonious, complete, and interpretable.

8. Stakeholders involved in promoting AI-based mental healthcare services

It is crucial for all stakeholders involved to advance and advocate for AI-based mental health treatments. Stakeholders who can aid in promoting these services include the following.

1. Mental and physical healthcare professionals (professionals in mental health, including psychiatrists, psychologists, psychiatric social workers, psychiatric nurses, and others, general healthcare workers, that is, doctors, nurses, and social workers).
2. Individuals suffering from mental health conditions, their families, and other caregivers.
3. Community-based groups, nongovernmental organizations (NGOs), social groups, and different welfare societies.
4. International, national, regional, and local print and social media
5. Numerous international, national, and local funding agencies.
6. International and national policy-making bodies include the World Health Organization, numerous national ministries, and parliamentary committees.

The reason behind involving a varied group of stakeholders is rooted in the notion that every stakeholder has an important role in using their knowledge to implement AI-based mental health services on a wider level. The major role and responsibilities of stakeholders in promoting AI-based mental healthcare services are.

1. **Covering AI-based mental health services in insurance policies:** The insurance policies and coverage do not cover mental healthcare. Even if the government passed the rules implementing these services in insurance, patients did not gain anything from it, and it will create a financial burden on them. Therefore, stakeholders and policymakers must pay attention to this element as well because insurance services must cover both general mental health treatments and AI-based mental health services.
2. **Improving Internet accessibility:** Due to the shortage of experienced and trained specialists, AI-based mental health treatments are primarily advantageous to the rural population and those living in remote locations. Although AI-based mental health services have many benefits, they are only useful when there is strong internet access. Experts and users cannot access these services without high-speed Internet. Therefore, this is an area that needs to be addressed by the government, organizations, policymakers, and stakeholders.
3. **Making AI-based services user-friendly:** For elderly people, using AI-based mental healthcare services can be tough due to the lack of technological knowledge. Additional problem is that AI-based mental health services are difficult to use in different cultural settings. Therefore, NGOs, social workers, and other stakeholders must assist this marginalized population in receiving AI-based mental health services.
4. **Regulating data breach:** AI-based mental healthcare services are advantageous, but it is also risky in certain ways. As these services and mobile apps proliferate quickly, clients may give their personal information in exchange for the services. The privacy of the clients might be jeopardized in routine data exchange between the developer and outside agencies (Martinez-Martin et al., 2020). Hence, governmental authorities, ministries, and other organizations must control this data breach and make these services more trustworthy.
5. **Providing funding and resources:** Another crucial role of the stakeholders in promoting AI-based mental health services is to provide adequate funding and resources to implement these services.
6. **Other roles and responsibilities of the stakeholders:** Other roles and responsibilities of the stakeholders are promoting AI-based mental health services, eliminating cultural barriers, project scheduling, tracking the progress, improving accessibility, and updating the services on time.

9. Steps to ethics-based governance of AI in mental healthcare services

Stakeholders can take several steps to operationalize AI in mental health services:

Phase 1: Exploratory phase

1. Conducting pilot research on the need for AI-based mental health services.
2. Syncing up professionals to start these services.
3. Reviewing content and supporting data.
4. Make an effort to guarantee fairness in accessing these services.

Phase 2: Phase of planning and execution

1. Establishing a business affiliate deal to limit data consumption.
2. Creating a guideline for using AI-based services.
3. Making fair and transparent data privacy, security, and uses.
4. Providing onboarding support.
5. Providing training and supervision for implementing these services.
6. Describing strategies for resolving safety and privacy issues.
7. Allocate sufficient funding.
8. Adopting AI-based services that are effective.
9. Establishing an appropriate referral procedure for the high-risk patient.
10. Making technical support accessible.
11. Ensuring the competence of AI-based services.
12. Conduct AI-based services on a small sample to test their efficacy.
13. Monitor the time and resources used to implement AI-based services.
14. Establish learning collaborations to exchange resources and knowledge.

Phase 3: Phase of maintenance

1. Throughout the period, improve the technology and implementation strategies.
2. Evaluation of shifting requirements and preferences across time (Naik et al., 2022).

10. Current trends and status of human-centric AI-based mental health services

As per Lavrentyeva (2022), AI-based mental health services have raised more than five billion dollars globally in the year 2021, an increase of about 139% from the year before. Human-centric AI concentrates on algorithms that are designed through human inputs and interaction. Human-centric

AI covers the gap between humans and AI with the perspective of understanding human emotions and feelings effectively. Currently, in general, in a healthcare setting, human-centric AI has been used in various ways such as health monitoring apps (Chekroud et al., 2021), Electronic healthcare records (Ghassemi et al., 2019), wearables (Mohr et al., 2017), image-based diagnosis (Arbabshirani et al., 2017), and AI-based health assessment and treatment (McIntosh et al., 2016).

In the past few years, human-centric AI services are also used in the mental health sector as well (Chekroud et al., 2021; Shatte et al., 2019). Initially, artificial intelligence services in mental healthcare settings were used in risk detection (DeMasi & Recht, 2017), diagnosis (Nobles et al., 2018), understanding prognosis (Nguyen et al., 2015), conducting self-assessment, monitoring symptoms, and tracking progress. The human-centric AI services in mental health are now being used in emotional intelligence-based chatbot services such as the Wysa app (emotional intelligence-based chatbot) (Fitzpatrick et al., 2017), Woebot app (Emotional intelligence-based CBT delivering app) (Inkster et al., 2018; Park et al., 2021), emotional intelligence-based stress management app (Mohr et al., 2017), and I-CBT apps.

11. Discussion and future implications

The objective of the current work is to understand the ethical consideration in mental healthcare through artificial intelligence. The healthcare sector is increasingly using AI, and the mental healthcare sector is currently adjusting to this development. Artificial intelligence-based mental healthcare services are in demand because of the scarcity of mental health specialists, the difficulty in accessing these services, and the sharp increase in mental health concerns. Although there are many benefits to AI based mental healthcare, there are significant ethical issues as well.

To use AI in a righteous way, it is important that algorithms used in detecting psychiatric illnesses must be accurate and must not increase any risk for clients. The gap in AI literacy must cover the literacy gap between clients and clinicians. However, there are not set standards for the use of AI in mental health services (Nebeker et al., 2017).

Secondly, the privacy concern: Services for mental healthcare using AI are freely available, but people need to check these service providers' privacy policies, app functionality, and discrimination policies.

Thirdly, mental healthcare experts must have sufficient training to provide AI-based mental healthcare services because it can help in reducing human biases.

Fourth, individuals do not need to completely rely on AI-based technology. These are helpful tools, but these cannot completely replace face-to-face treatment services, in-patient services, and other important aspects of mental health treatment.

Fifth, numerous AI apps are made to give users information without being directly accessible by a person. While the user may benefit from this, there are also drawbacks. Humans are sociable creatures who frequently gain from interacting with others. Utilizing AI apps exclusively can decrease the chances of social contact. This may be an issue, particularly for those who lack social abilities or are isolated and lonely. In contrast, certain AI applications are made to give feedback automatically. Although some individuals might find this useful, others might find it harmful. Human feedback is frequently valuable because it enables users to learn from their errors and alter their behavior as they go. Without human input, a user would not be aware of their errors and may not have the chance to do better next time.

Lastly, an expert authority doesn't control applications either. Because of this, people who use them for mental health services frequently do not receive the same level of care that would be offered by a qualified mental health professional. Instead, consumers might be utilizing a program built by programmers that have not undergone the same level of scrutiny as a mental health expert would. People do not have the same legal safeguards when using mental health apps as they do when they visit a qualified mental health expert. The argument about whether artificial intelligence or the Internet of Things can take over the human element is never-ending. While the concept of automation of mental health services has found advocacy in larger sections of society, there are critiques for the involvement of AI in psychotherapy and psychiatric services owing to the idiosyncratic approach followed during therapy and consultation. The feelings of compassion, empathy, and unconditional positive regard cannot be offered by artificial intelligence; however, its easy accessibility and wide approach cannot be overlooked. Human elements can employ artificial intelligence through effective monitoring and supervision. However, the welfare of clients/patients is the topmost priority and as in the case of in-person counseling/consultation, artificial intelligence must also keep up with ethical guidelines.

To understand the efficacy of artificial intelligence for mental healthcare, there is a need for high-quality quantitative and qualitative research. Future research can focus on the effectiveness of clinical services provided by AI to clients. Secondly, researchers can focus on the role of clinicians if the AI will deliver services for mental health treatment. Thirdly, future research can also assess the issues of psychotherapeutic alliances and how it relates to online Internet-based therapy (Özdemir, 2019).

12. Conclusion

The usage of AI in healthcare services is rapidly increasing. Successful AI use in mental healthcare services might significantly raise the standard of treatment. New techniques for diagnosing psychiatric illnesses, observation, monitoring, and treatment can improve therapeutic outcomes and rebalance the workload of clinicians. There is a lot of potentials, but there are also lots of risks and difficulties. To ensure the successful application of this new technology, these will need to be navigated carefully. Collaboration among various stakeholders like mental health experts, scientists, clinicians, and other experts is required to use AI to its full potential in the field of mental health care. In the field of mental health treatment, AI has a bright future. As a specialist in mental healthcare, we need to take a proactive spirit using AI to improve our clinical experience and learning, helping to transform mental healthcare practices, and improving clients' health.

References

Arbabshirani, M. R., Plis, S., Sui, J., & Calhoun, V. D. (2017). Single subject prediction of brain disorders in neuroimaging: Promises and pitfalls. *NeuroImage, 145*, 137–165. https://doi.org/10.1016/j.neuroimage.2016.02.079

Blease, C., Locher, C., Leon-Carlyle, M., & Doraiswamy, M. (2020). Artificial intelligence and the future of psychiatry: Qualitative findings from a global physician survey. *Digital Health, 6*, Article 205520762096835. https://doi.org/10.1177/2055207620968355

Bzdok, D., & Meyer-Lindenberg, A. (2018). Machine learning for precision psychiatry: Opportunities and challenges. *Biological Psychiatry: Cognitive Neuroscience and Neuroimaging, 3*(3), 223–230. https://doi.org/10.1016/j.bpsc.2017.11.007

Bzdok, D., & Yeo, B. T. (2017). Inference in the age of big data: Future perspectives on neuroscience. *NeuroImage, 155*, 549–564. https://doi.org/10.1016/j.neuroimage.2017.04.061

Castro, V. M., Roberson, A. M., McCoy, T. H., Wiste, A., Cagan, A., Smoller, J. W., Rosenbaum, J. F., Ostacher, M., & Perlis, R. H. (2015). Stratifying risk for renal insufficiency among lithium-treated patients: An electronic health record study. *Neuropsychopharmacology, 41*(4), 1138–1143. https://doi.org/10.1038/npp.2015.254

Chekroud, A. M., Bondar, J., Delgadillo, J., Doherty, G., Wasil, A., Fokkema, M., Cohen, Z., Belgrave, D., DeRubeis, R., Iniesta, R., Dwyer, D., & Choi, K. (2021). The promise of machine learning in predicting treatment outcomes in psychiatry. *World Psychiatry, 20*(2), 154–170. https://doi.org/10.1002/wps.20882

Craig, T. K., Rus-Calafell, M., Ward, T., Leff, J. P., Huckvale, M., Howarth, E., Emsley, R., & Garety, P. A. (2018). AVATAR therapy for auditory verbal hallucinations in people with psychosis: A single-blind, randomized controlled trial. *The Lancet Psychiatry, 5*(1), 31–40. https://doi.org/10.1016/S2215-0366(17)30427-3

Dattani, S., Ritchie, H., & Roser, M. (2018). Mental Health. Published online at OurWorldInData.org. Retrieved from https://ourworldindata.org/mental-health. Accessed on December, 29th, 2022.

DeMasi, O., & Recht, B. (2017). A step towards quantifying when an algorithm can and cannot predict an individual's well-being. In *Proceedings of the 2017 ACM international joint conference on pervasive and ubiquitous computing and proceedings of the 2017 ACM international symposium on wearable computers*. https://doi.org/10.1145/3123024.3125609

Durstewitz, D., Koppe, G., & Meyer-Lindenberg, A. (2019). Deep neural networks in psychiatry. *Molecular Psychiatry, 24*(11), 1583–1598. https://doi.org/10.1038/s41380-019-0365-9

Dwyer, D. B., Falkai, P., & Koutsouleris, N. (2018). Machine learning approaches for clinical psychology and psychiatry. *Annual Review of Clinical Psychology, 14*(1), 91–118. https://doi.org/10.1146/annurev-clinpsy-032816-045037

Evans, R. S. (2016). Electronic health records: Then, now, and in the future. *Yearbook of Medical Informatics, 25*(S 01), S48–S61. https://doi.org/10.15265/iys-2016-s006

Fitzpatrick, K. K., Darcy, A., & Vierhile, M. (2017). Delivering cognitive behavior therapy to young adults with symptoms of depression and anxiety using a fully automated conversational agent (Woebot): A randomized controlled trial. *JMIR Mental Health, 4*(2), e19. https://doi.org/10.2196/mental.7785

Fulmer, R., Joerin, A., Gentile, B., Lakerink, L., & Rauws, M. (2021). Using psychological artificial intelligence (Tess) to relieve symptoms of depression and anxiety: Randomized controlled trial. *JMIR Mental Health, 5*(4), e64. https://doi.org/10.2196/mental.9782

Fusar-Poli, P., Hijazi, Z., Stahl, D., & Steyerberg, E. W. (2018). The science of prognosis in psychiatry. *JAMA Psychiatry, 75*(12), 1289. https://doi.org/10.1001/jamapsychiatry.2018.2530

Gabbard, G. O., & Crisp-Han, H. (2016). The early career psychiatrist and the psychotherapeutic identity. *Academic Psychiatry, 41*(1), 30–34. https://doi.org/10.1007/s40596-016-0627-7

Garg, M. (2023). Mental health analysis in social media posts: A survey. *Archives of Computational Methods in Engineering*, 1–24.

Garg, K., Kumar, C. N., & Chandra, P. S. (2019). Number of psychiatrists in India: Baby steps forward, but a long way to go. *Indian Journal of Psychiatry, 61*(1), 104–105. https://doi.org/10.4103/psychiatry.IndianJPsychiatry_7_18

Ghassemi, M., Naumann, T., Schulam, P., Beam, A. L., Chen, I. Y., & Ranganath, R. (2019). Practical guidance on artificial intelligence for health-care data. *The Lancet Digital Health, 1*(4), e157–e159. https://doi.org/10.1016/s2589-7500(19)30084-6

Graham, S., Depp, C., Lee, E. E., Nebeker, C., Tu, X., Kim, H., & Jeste, D. V. (2019). Artificial intelligence for mental health and mental illnesses: An overview. *Current Psychiatry Reports, 21*(11). https://doi.org/10.1007/s11920-019-1094-0

Hariharan, S., Krishna Prasanth, B., Stephen, T., & Aljin, V. (2020). Mental health scenario in India. *Annals of Medical and Health Sciences Research, 10*(1), 1058–1059.

Hengstler, M., Enkel, E., & Duelli, S. (2016). Applied artificial intelligence and trust—the case of autonomous vehicles and medical assistance devices. *Technological Forecasting and Social Change, 105*, 105–120. https://doi.org/10.1016/j.techfore.2015.12.014

Inkster, B., Sarda, S., & Subramanian, V. (2018). An empathy-driven, conversational artificial intelligence agent (Wysa) for digital mental well-being: Real-world data evaluation mixed-methods study. *JMIR mHealth and uHealth, 6*(11), Article e12106. https://doi.org/10.2196/12106

Janssen, R. J., Mourão-Miranda, J., & Schnack, H. G. (2018). Making individual prognoses in psychiatry using neuroimaging and machine learning. *Biological Psychiatry: Cognitive Neuroscience and Neuroimaging, 3*(9), 798–808. https://doi.org/10.1016/j.bpsc.2018.04.004

Jeste, D. V., Glorioso, D., Lee, E. E., Daly, R., Graham, S., Liu, J., Paredes, A. M., Nebeker, C., Tu, X. M., Twamley, E. W., Van Patten, R., Yamada, Y., Depp, C., & Kim, H. (2019). Study of independent living residents of a continuing care senior housing community: Sociodemographic and clinical associations of cognitive, physical, and mental health. *American Journal of Geriatric Psychiatry, 27*(9), 895–907. https://doi.org/10.1016/j.jagp.2019.04.002

Jeste, D. V., Lee, E. E., & Cacioppo, S. (2020). Battling the modern behavioral epidemic of loneliness. *JAMA Psychiatry,* 77(6), 553. https://doi.org/10.1001/jamapsychiatry.2020.0027

Jiang, F., Jiang, Y., Zhi, H., Dong, Y., Li, H., Ma, S., Wang, Y., Dong, Q., Shen, H., & Wang, Y. (2017). Artificial intelligence in healthcare: Past, present and future. *Stroke and Vascular Neurology, 2*(4), 230–243. https://doi.org/10.1136/svn-2017-000101

Kaur, S., Bhardwaj, R., Jain, A., Garg, M., & Saxena, C. (December 2022). Causal categorization of mental health posts using transformers. In *Proceedings of the 14th annual meeting of the forum for information retrieval evaluation* (pp. 43–46).

Kloppel, S., Stonnington, C. M., Barnes, J., Chen, F., Chu, C., Good, C. D., Mader, I., Mitchell, L. A., Patel, A. C., Roberts, C. C., Fox, N. C., Jack, C. R., Ashburner, J., & Frackowiak, R. S. (2008). Accuracy of dementia diagnosis–a direct comparison between radiologists and a computerized method. *Brain, 131*(11), 2969–2974. https://doi.org/10.1093/brain/awn239

Koleck, T. A., Dreisbach, C., Bourne, P. E., & Bakken, S. (2019). Natural language processing of symptoms documented in free-text narratives of electronic health records: A systematic review. *Journal of the American Medical Informatics Association, 26*(4), 364–379. https://doi.org/10.1093/jamia/ocy173

Lavrentyeva, Y. (2022). *AI in mental health - examples, benefits and trends — ITRex*. ITRex. https://itrexgroup.com/blog/ai-mental-health-examples-trends/.

Luxton, D. D. (2014). Artificial intelligence in psychological practice: Current and future applications and implications. *Professional Psychology: Research and Practice, 45*(5), 332–339. https://doi.org/10.1037/a0034559

Mahajan, S. (2020). *Artificial intelligence in healthcare market-global industry analysis, size and growth report* (p. 2026).

Martinez-Martin, N., Dasgupta, I., Carter, A., Chandler, J. A., Kellmeyer, P., Kreitmair, K., Weiss, A., & Cabrera, L. Y. (2020). Ethics of digital mental health during COVID-19: Crisis and opportunities. *JMIR Mental Health,* 7(12), Article e23776. https://doi.org/10.2196/23776

Maslow, A. H. (1943). A theory of human motivation. *Psychological Review, 50*(4), 370–396.

McCarthy, J. (1989). Artificial intelligence, logic, and formalizing common sense. In R. H. Thomason (Ed.), *Philosophical logic and artificial intelligence*. Dordrecht: Springer. https://doi.org/10.1007/978-94-009-2448-2_6

McIntosh, A. M., Stewart, R., John, A., Smith, D. J., Davis, K., Sudlow, C., Corvin, A., Nicodemus, K. K., Kingdon, D., Hassan, L., Hotopf, M., Lawrie, S. M., Russ, T. C., Geddes, J. R., Wolpert, M., Wölbert, E., & Porteous, D. J. (2016). Data science for mental health: A UK perspective on a global challenge. *The Lancet Psychiatry, 3*(10), 993–998. https://doi.org/10.1016/s2215-0366(16)30089-x

Miller, D. D., & Brown, E. W. (2018). Artificial intelligence in medical practice: The question to the answer? *The American Journal of Medicine, 131*(2), 129–133. https://doi.org/10.1016/j.amjmed.2017.10.035

Mohr, D. C., Zhang, M., & Schueller, S. M. (2017). Personal sensing: Understanding mental health using ubiquitous sensors and machine learning. *Annual Review of Clinical Psychology, 13*(1), 23–47. https://doi.org/10.1146/annurev-clinpsy-032816-044949

Mörch, C. M., Gupta, A., & Mishara, B. L. (2019). *Canada protocol: An ethical checklist for the use of artificial Intelligence in suicide prevention and mental health.* arXiv preprint arXiv: 1907.07493.

Morley, J., & Floridi, L. (2020). The limits of empowerment: How to reframe the role of mHealth tools in the healthcare ecosystem. *Science and Engineering Ethics, 26*(3), 1159–1183.

Mwangi, B., Wu, M., Cao, B., Passos, I. C., Lavagnino, L., Keser, Z., Zunta-Soares, G. B., Hasan, K. M., Kapczinski, F., & Soares, J. C. (2016). Individualized prediction and clinical staging of bipolar disorders using neuroanatomical biomarkers. *Biological Psychiatry: Cognitive Neuroscience and Neuroimaging, 1*(2), 186–194. https://doi.org/10.1016/j.bpsc.2016.01.001

Naik, N., Hameed, B. M., Shetty, D. K., Swain, D., Shah, M., Paul, R., Aggarwal, K., Ibrahim, S., Patil, V., Smriti, K., Shetty, S., Rai, B. P., Chlosta, P., & Somani, B. K. (2022). Legal and ethical consideration in artificial intelligence in healthcare: Who takes responsibility? *Frontiers in Surgery, 9*. https://doi.org/10.3389/fsurg.2022.862322

Nebeker, C., Harlow, J., Espinoza Giacinto, R., Orozco-Linares, R., Bloss, C. S., & Weibel, N. (2017). Ethical and regulatory challenges of research using pervasive sensing and other emerging technologies: IRB perspectives. *AJOB Empirical Bioethics, 8*(4), 266–276. https://doi.org/10.1080/23294515.2017.1403980

Nguyen, T., O'Dea, B., Larsen, M., Phung, D., Venkatesh, S., & Christensen, H. (2015). Using linguistic and topic analysis to classify sub-groups of online depression communities. *Multimedia Tools and Applications, 76*(8), 10653–10676. https://doi.org/10.1007/s11042-015-3128-x

Nobles, A. L., Glenn, J. J., Kowsari, K., Teachman, B. A., & Barnes, L. E. (2018). Identification of imminent suicide risk among young adults using text messages. In *Proceedings of the 2018 CHI conference on human factors in computing systems*. https://doi.org/10.1145/3173574.3173987

Özdemir, V. (2019). Not all intelligence is artificial: Data science, automation, and AI meet HI. *OMICS: A Journal of Integrative Biology, 23*(2), 67–69. https://doi.org/10.1089/omi.2019.0003

Pang, Z., Yuan, H., Zhang, Y. T., & Packirisamy, M. (2018). Guest editorial health engineering is driven by industry 4.0 for an aging society. *IEEE Journal of Biomedical and Health Informatics, 22*(6), 1709–1710. https://doi.org/10.1109/jbhi.2018.2874081

Park, J., Cho, H., Cha, J. M., Kim, J. H., Yoo, S., Kim, H., & Cha, J. (2019). P1-270: Machine learning prediction of future incidence of Alzheimer's disease using population-wide electronic health records. *Alzheimer's and Dementia, 15*, P342–P343. https://doi.org/10.1016/j.jalz.2019.06.825

Park, S., Thieme, A., Han, J., Lee, S., Rhee, W., & Suh, B. (2021). "I wrote as if I were telling a story to someone I knew.": Designing chatbot interactions for expressive writing in mental health. In *Designing interactive systems conference 2021*. https://doi.org/10.1145/3461778.3462143

Passos, I. C., Ballester, P. L., Barros, R. C., Librenza-Garcia, D., Mwangi, B., Birmaher, B., Brietzke, E., Hajek, T., Lopez Jaramillo, C., Mansur, R. B., Alda, M., Haarman, B. C., Isometsa, E., Lam, R. W., McIntyre, R. S., Minuzzi, L., Kessing, L. V., Yatham, L. N., Duffy, A., & Kapczinski, F. (2019). Machine learning and big data analytics in bipolar disorder: A position paper from the international society for bipolar disorders big data task force. *Bipolar Disorders, 21*(7), 582–594. https://doi.org/10.1111/bdi.12828

Passos, I. C., & Mwangi, B. (2018). Machine learning-guided intervention trials to predict treatment response at an individual patient level: An important second step following randomized clinical trials. *Molecular Psychiatry, 25*(4), 701–702. https://doi.org/10.1038/s41380-018-0250-y

Redlich, R., Almeida, J. J., Grotegerd, D., Opel, N., Kugel, H., Heindel, W., Arolt, V., Phillips, M. L., & Dannlowski, U. (2014). Brain Morphometric biomarkers distinguishing unipolar and bipolar depression. *JAMA Psychiatry, 71*(11), 1222. https://doi.org/10.1001/jamapsychiatry.2014.1100

Rogers, C. R. (1957). The necessary and sufficient conditions of therapeutic personality change. *Journal of Consulting Psychology, 21*(2), 95–103. https://doi.org/10.1037/h0045357

Rubeis, G. (2022). iHealth: The ethics of artificial intelligence and big data in mental healthcare. *Internet Interventions, 28,* Article 100518. https://doi.org/10.1016/j.invent.2022.100518

Scheffler, R. M., Bruckner, T., Fulton, B. D., Yoon, J., Shen, G., & Chisholm, D. (2011). Human resources for mental health: Workforce shortages in low- and middle-income countries. *Human Resources for Health Observer, 11,* VII–VIII. ISBN 978-92-4-150101-9.

Schmaal, L., Marquand, A. F., Rhebergen, D., Van Tol, M., Ruhé, H. G., Van der Wee, N. J., Veltman, D. J., & Penninx, B. W. (2015). Predicting the naturalistic course of major depressive disorder using clinical and multimodal neuroimaging information: A multivariate pattern recognition study. *Biological Psychiatry, 78*(4), 278–286. https://doi.org/10.1016/j.biopsych.2014.11.018

Schwab, K. (2017). *The fourth industrial revolution* (1st ed.). Currency.

Shatte, A. B., Hutchinson, D. M., & Teague, S. J. (2019). Machine learning in mental health: A scoping review of methods and applications. *Psychological Medicine, 49*(09), 1426–1448. https://doi.org/10.1017/s0033291719000151

Sheikhalishahi, S., Miotto, R., Dudley, J. T., Lavelli, A., Rinaldi, F., & Osmani, V. (2019). Natural language processing of clinical notes on chronic diseases: Systematic review. *JMIR Medical Informatics,* 7(2), Article e12239. https://doi.org/10.2196/12239

Shen, N., Sequeira, L., Silver, M. P., Carter-Langford, A., Strauss, J., & Wiljer, D. (2019). Patient privacy perspectives on health information exchange in a mental health context: Qualitative study. *JMIR Mental Health, 6*(11), Article e13306. https://doi.org/10.2196/13306

Srivastava, K., Chatterjee, K., & Bhat, P. S. (2016). Mental health awareness: The Indian scenario. *Industrial Psychiatry Journal, 25*(2), 131–134. https://doi.org/10.4103/ipj.ipj_45_17

Tran, T., Luo, W., Phung, D., Harvey, R., Berk, M., Kennedy, R. L., & Venkatesh, S. (2014). Risk stratification using data from electronic medical records better predicts suicide risks than clinician assessments. *BMC Psychiatry, 14*(1). https://doi.org/10.1186/1471-244x-14-76

Turing, A. M. (2007). *Computing machinery and intelligence A.M. Turing* (pp. 433–460). Available from https://linkinghub.elsevier.com/retrieve/pii/B978012386980750023X.

Vieira, S., Pinaya, W. H., & Mechelli, A. (2017). Using deep learning to investigate the neuroimaging correlates of psychiatric and neurological disorders: Methods and applications. *Neuroscience and Biobehavioral Reviews, 74,* 58–75. https://doi.org/10.1016/j.neubiorev.2017.01.002

Vigo, D., Thornicroft, G., & Atun, R. (2016). Estimating the true global burden of mental illness. *The Lancet Psychiatry, 3*(2), 171–178. https://doi.org/10.1016/s2215-0366(15)00505-2

Vilaza, G. N., & McCashin, D. (2021). Is the automation of digital mental health ethical? Applying an ethical framework to chatbots for cognitive behavior therapy. *Frontiers in Digital Health, 3,* 689–736. https://doi.org/10.3389/fdgth.2021.689736

Vos, T., Abajobir, A. A., Abate, K. H., Abbafati, C., Abbas, K. M., Abd-Allah, F., Abdulkader, R. S., Abdulle, A. M., Abebo, T. A., Abera, S. F., Aboyans, V., Abu-Raddad, L. J., Ackerman, I. N., Adamu, A. A., Adetokunboh, O., Afarideh, M., Afshin, A., Agarwal, S. K., Agarwal, R., & Murray, C. L. J. (2017). Global, regional, and national incidence, prevalence, and years lived with disability for 328 diseases and injuries for 195 countries, 1990-2016: A systematic analysis for the global burden of disease study 2016. *The Lancet, 390,* 1211–1259. https://doi.org/10.1016/S0140-6736(17)32154-2

World Health Organization. (2018). *Mental health in India.*

CHAPTER ELEVEN

Human AI: Social network analysis

Umesh Gupta[1], Gargi Trivedi[2] and Divya Singh[2]
[1]SR University, Warangal, Telangana, India
[2]Bennett University, Greater Noida, Uttar Pradesh, India

1. Introduction to social network analysis

Social network analysis (SNA) originated in early sociological research, emphasizing the study of relationship patterns between social actors. In essence, SNA is an approach for investigating social structures, which is why SNA is often referred to as structural analysis. It is widely used in various fields like sociology, psychology, marketing, and management to understand the relationships and dynamics within complex networks (SNA Methods and Techniques, 2023; Graph Theory and Applications, 2023; Incidence Matrix, 2023). In today's world, social media platforms such as Facebook, Twitter, and LinkedIn have become famous avenues for social networking, creating a vast network of connections among billions of users. As much of this networking takes place in virtual spaces, social scientists and computer scientists have many opportunities and challenges for research (SNA Methods and Techniques, 2023). While the global interconnectivity of social networks has positive effects, it also has negative consequences, such as the spreading of hate and intolerance by local and global terrorist groups. Understanding the dynamics of social networks is crucial for improving social conditions and developing effective interventions. The concept of social networking, which refers to building and maintaining social relationships, has been a part of human life for a long time (SNA and Its Use Case, 2023; Graph Theory and Applications, 2023). However, the modern version of social networking that we know today emerged with the Internet and web-based technologies. In the 1990s, online communities and chat rooms became popular, enabling people to connect and communicate over the Internet. The first social networking site, Six Degrees, was launched in 1997, allowing users to create profiles and connect with others.

Other social networking sites followed, such as Friendster (2002), Myspace (2003), and eventually, Facebook (2004), which became the dominant player in the field (Knoke & Yang, 2020/2020; SNA Use Case, 2023;

Emotional AI and Human-AI Interactions in Social Networking
ISBN: 978-0-443-19096-4
https://doi.org/10.1016/B978-0-443-19096-4.00004-3

Intro to SNA, 2023). These sites enabled users to connect with others, share information and media, and create online communities centered around shared interests and affiliations. The rise of smartphones and mobile devices has further accelerated the growth of social networking, with platforms like Twitter, Instagram, Snapchat, and TikTok providing new ways for people to connect and share information. Today, social networking is ubiquitous, with billions worldwide using social media and other platforms to connect daily (Prell, 2012/2012; Weighted Decision Matrix, 2023/2023; SNA Applications, 2023; Reachability Matrix, 2023). The structure-based role of network performers is plotted in Fig. 11.1.

The basic steps involved in the working of SNA are as follows.

1. Data collection is the initial step in collecting data about the relationships and interactions between individuals within a network (Data Collection Guide, 2023; Centrality Measure, 2023; Social Network Analysis, 2023; Wasserman & Faust, 1999/1999; Electronic Communication Network, 2023). The data can be managed through surveys, observation, or digital tools like social media. Data collection is a crucial step in SNA, as the quality and accuracy of the study depend on the data gathered (Data Collection Methods, 2023). Several methods can be used to collect data for SNA, including:
   a. *Surveys:* Generally, these are used to collect information about the relationships and interactions between individuals within a network

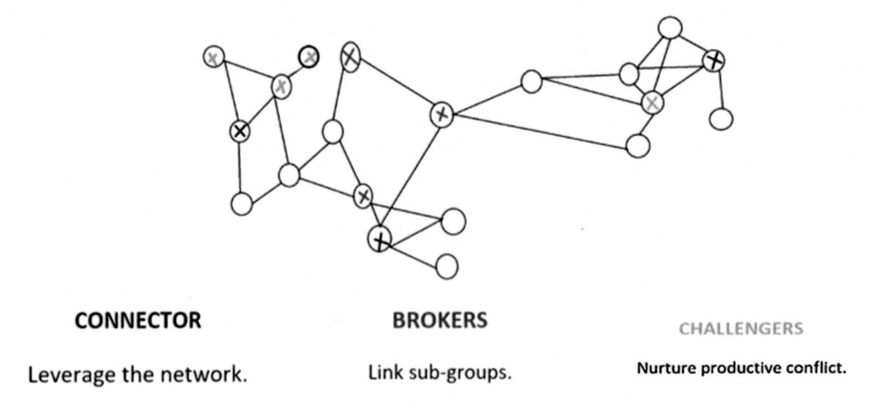

Figure 11.1 Structure-based role of network performers.

(Data Collection Methods, 2023). The survey can be a questionnaire or interview conducted online or in person.

b. *Observation:* Observation is another method that can be used to collect data for SNA. Researchers can observe the interactions and relationships between individuals within a network in a natural setting, such as in a workplace or social gathering (Data Collection Methods, 2023).

c. *Digital tools:* Taking the example of the most renowned platforms, such as Facebook, Twitter, and LinkedIn, can be used to collect data for SNA (Data Collection Methods, 2023; Electronic Communication Network, 2023). This platform allows researchers to gather information about the relationships and interactions between individuals within a network. People use such platforms to share things related to their personal life or their present happening (Garg, 2022).

d. *Electronic communication data:* Email, instant messaging, and other forms of electronic communication can be analyzed to collect data for SNA (What is Data Collection, 2023). This method allows researchers to gather information about the frequency and content of communication between individuals within a network (Data Collection Methods, 2023; What is Data Collection, 2023; Electronic Communication Network, 2023).

e. *Historical records:* Historical records, such as meeting minutes and agendas, can be used to collect data for SNA (What is Data Collection, 2023; Reachability Matrix, 2023). This method allows researchers to gather information about individual relationships and interactions over time. It is important to note that data collection for SNA can be challenging and time-consuming. Remembering that the collected data are precise, complete, and resembles the analyzed network (Data Collection Methods, 2023; Weighted Decision Matrix, 2023/2023; What is Data Collection, 2023; Electronic Communication Network, 2023) is essential.

f. The type of collected data: The data collected for SNA depend on the research question and the network's context. However, the following are the most common types of data collected for SNA (Creating Dynamic Network Views, 2023):

- *Relationships:* The primary data collected for SNA are information about the relationships between individuals within a network. This information can include the type of relationship (e.g., friendship, collaboration, etc.), the relationship's strength, and the interaction frequency (TCP/IP, 2023).

- *Demographic information:* Information about the individuals within the network, such as age, gender, education level, occupation, etc., can also be collected for SNA (TCP/IP, 2023). This information will help you figure out what kind of people are in the network and how they relate to and interact with each other (Creating Dynamic Network Views, 2023; TCP/IP, 2023).
- *Communication data*: Information about the frequency and content of communication between individuals within a network can also be collected for SNA (TCP/IP, 2023). This information can be gathered through email, instant messaging, or other forms of electronic communication (Creating Dynamic Network Views, 2023; TCP/IP, 2023).
- *Social media data:* The most renowned platforms, such as Facebook, Twitter, and LinkedIn, can be used to collect data for SNA. These data can include information about the relationships between individuals, the frequency and content of communication, and the activity of individuals within the network (TCP/IP Model, 2023; Tcp/Ip, 2023).
- *Network structure:* Information about the design of the network, such as the number of individuals, the number of connections, and the distribution of relationships, can also be collected for SNA. This information can help to understand the overall structure of the network and identify critical features, such as centrality and clustering (Creating Dynamic Network Views, 2023; TCP/IP Model, 2023; Tcp/Ip, 2023). The data collected for SNA should represent the network being analyzed and provide enough information to answer the research questions being addressed. It is mandatory to ensure that the data collected are accurate, complete, and consistent, as this will affect the validity of the results obtained from the analysis (Creating Dynamic Network Views, 2023; TCP/IP Model, 2023; Tcp/Ip, 2023).

2. Data representation: The collected data are then represented using graph theory, where individuals are represented as nodes and relationships are defined as edges connecting the nodes (Creating Dynamic Network Views, 2023; TCP/IP Model, 2023; Tcp/Ip, 2023). Data presentation is an essential step in SNA because it visually represents relationships and interactions in the network (TCP/IP Model, 2023; Network Traffic

Analyser, 2023; Wang et al., 2016). The most common ways to express data in SNA are:

a. *Edge node representation:* Among all the methods, this is the most common representation used in SNA. In this method, people in the network are represented by nodes, and relationships between people are defined by edges connecting the nodes. Various graphical elements, such as circles and lines, can visualize a network representing nodes and edges (VPN and Firewall, 2023; Wang et al., 2016).

b. *Matrix representation:* In this method, networks are represented as matrices, and individuals are defined as rows and columns. The cells of a matrix represent relationships between people, and the values in cells can be binary numbers (such as 0 or 1) or weights (such as the strength of the connection). Graphical representation: This technique uses various visual elements, such as graphs and charts, to represent network data. Graphical models can describe different aspects of networks, such as structure, dynamics, and relationships between people (VPN and Firewall, 2023; Wang et al., 2016; Encryption Algorithm, 2023).

c. *Dynamic view:* This technique shows changes in the network over time. This method visualizes the network as a series of snapshots, allowing researchers to see how the network evolves (TCP/IP Model, 2023). Displaying data in SNA is crucial as it helps to make analysis results more accessible and understandable. Choosing a representation method that clearly and accurately represents the network is vital. This is because it affects the validity of the results obtained from the analysis. The presentation method should be selected according to the research question being addressed and the data being analyzed (Bober & Kaproń, 2010; Cook et al., 1983; Encryption Algorithm, 2023).

3. Network analysis: In this step, various metrics and algorithms are used to analyze the network data and identify critical features like centrality, degree, clustering coefficient, etc. These metrics help to understand the structure and dynamics of the network. We will investigate it further (Bober & Kaproń, 2010; Cook et al., 1983; Encryption Algorithm, 2023).

4. Interpretation is crucial in SNA. It allows researchers to make sense of the data and draw insights about the relationships and interactions within a network. The results obtained from the network analysis are then interpreted to draw insights into the network's relationships and interactions.

This can help to identify the most influential individuals, communities, and patterns of behavior within the network. The following are some of the interpretations that can be obtained from network analysis:

- *Centrality:* Centrality is all about the significance of individuals within a network based on their connections to others. The most common centrality measures include degree centrality (number of links), betweenness centrality (number of times an individual acts as a bridge between other individuals), and eigenvector centrality (importance of an individual based on the importance of those they are connected to) (Bober & Kaproń, 2010; Valeri & Baggio, 2021a, 2021b).
- *Clustering:* Clustering measures the degree to which individuals within a network are connected, forming tightly knit groups. This can identify communities or sub-groups within the network (Valeri & Baggio, 2021a).
- *Structuring:* Network structure refers to the general arrangement of connections and relationships in a network. By analyzing network structure, you can understand connectivity and identify key characteristics such as centrality, clustering, and the presence of subgroups or communities (Valeri & Baggio, 2021b).
- *Dynamics:* Network dynamics describe changes in a network over time. Analyzing network dynamics lets you understand pattern changes and identify critical factors influencing network evolution (Valeri & Baggio, 2021b).
- *Influence:* Influence refers to people's effect on each other in a network. Impact analysis allows you to understand the flow of information and resources within your network and identify key individuals who shape your network. Interpretation is essential in SNA because it lets everyone gain insight into the relationships and interactions of networks and answer questions being addressed during their research. Interpretation of network analysis results should be based on sound statistical and analytical methods, and it is vital to ensure that the results are valid and reliable (Valeri & Baggio, 2021b).
- Visualization: The network analysis results are often visualized using tools like network graphs, matrices, and other visual representation methods to help make the findings more accessible and easier to understand (AlKheder et al., 2023; Valeri & Baggio, 2021a; Webber et al., 2023).

2. Understanding networking and their partners

The basic understanding of networking is connectivity. In technical terms, networking refers to connecting computers, devices, and other hardware components to facilitate communication and exchanging information and resources. It can refer to local area networks (LANs), wide area networks (WANs), and the Internet. At its core, networking involves connecting multiple devices, physically or wirelessly, and configuring them to exchange information reliably and efficiently (Yang & Knoke, 2001). This requires a combination of hardware components (such as switches, routers, and modems), software technologies (such as protocols and operating systems), and a set of established rules and standards (such as the IP and the transmission control protocol (TCP)) (Nooraie, 2021). In addition to allowing devices to communicate with each other, networks also provide several other benefits, including.

- Resource sharing: Networks allow users to share resources, such as printers and file servers, among multiple devices (Yousefi Nooraie et al., 2020).
- Increased accessibility: Networks allow users to access information and resources from any connected device rather than being limited to the resources available on a single machine.
- Enhanced connectivity: The network facilitates communication between devices, allowing users to email, chat, and participate in video conferences, among other things. Key concepts such as IP addresses, subnets, and routing are essential to understand how networks work. Familiarity with protocols and technologies such as TCP/IP, Ethernet, and Wi-Fi is also helpful. These basics allow you to explore more advanced topics such as network security, cloud computing, and network automation. Individuals and organizations comprise a network's nodes, while interconnections between nodes depict their affiliations or information flows (Nooraie, 2021). Organizational network analysis, or ONA, is the methodology management consultants employ with their customer base. SNA provides mathematical and visual evaluations of interconnections (Nooraie, 2021; Toomet, 2013).

2.1 Network's various roles and groupings

To understand networking and its users, we must have the collective data of the location and grouping of users in the network.

- *Degree centrality:* Technically, this talks about the significance of a node in a network utterly dependent on the number of connections made. It is often used to confirm dominant nodes in social networks or to measure the trustworthiness of a network (Buhai & van der Leij, 2023). The degree of centrality can be calculated for both directed and undirected networks and is based on the connection of nodes with numerous edges. In controlled networks, the node's centrality is the total number of incoming and outgoing links; in undirected networks, it is the sum of links. The centrality degree measures the node's direct influence on the local network (Kong et al., 2019). This is because nodes with the most incredible power of centrality usually have a more direct impact on the nodes they are connected. The degree of centrality can also be used to identify influential nodes that are part of a shorter distance between two nodes that connect a different network component.
- *Betweenness centrality:* Betweenness divides all shortest routes that pass via a given node in a graph. In other words, it measures the number of times a node connects the fastest ways between two different nodes. Mediation centrality is essential in network analysis because it helps identify the most critical nodes in a graph (Yoon et al., 2019). For example, nodes with the highest median centrality can be considered the most influential member. Similarly, nodes with high median centrality in transport networks can become virtual nodes in a social network. Compute centrality by mediation is calculating the closest path between any two nodes that pass via a given node for every pair of nodes in the graph. This calculation can be computationally intensive, especially for large graphs. However, betweenness centrality can be efficiently calculated using an algorithm such as the Brandeis algorithm (Buhai & van der Leij, 2023; Kong et al., 2019; Yoon et al., 2019).
- *Closeness Centrality:* A particular node's centrality in a network can be determined by its proximity centrality. The value of any two nodes in a network is the sum of the lengths of the shortest paths between them. High proximity centrality node impacts other nodes in the network. It evaluates a node's connectivity to every other node on the web (Yoon et al., 2019). One of the most used metrics of centrality in network analysis is proximity centrality, which is frequently used to identify essential participants in a network (SNA and Its Use Case, 2023; Yoon et al., 2019).
- *Network centralization* refers to the degree to which a network is centralized or decentralized. In SNA, centralization refers to the extent to

which a few nodes in a network hold a disproportionate amount of power or influence compared to other nodes. Various methods can be employed to evaluate the level of centralization within a network, such as degree centrality, betweenness centrality, and eigenvector centrality. Degree centrality determines the count of links a node has with other nodes in the network. Nodes with a high degree of centrality are often considered highly influential, as they have many connections to other nodes and can transmit information, resources, or influence throughout the network. Betweenness centrality is a metric used to assess the frequency with which a node functions as a mediator or bridge between other nodes in a network. Nodes with a significant betweenness centrality are typically considered crucial gatekeepers, as they regulate the movement of information and resources between different network parts. Eigenvector centrality measures the relative significance of a node based on the number and importance of the connections it has to other nodes in the network. The most influential nodes are those with the most remarkable eigenvector centrality (Majeed & Rauf, 2020). They are connected to other highly effective nodes and have a broader reach regarding information, resources, and influence. Overall, network centralization is an integral part of SNA as it provides insights into the structure and dynamics of a network and can help identify the most influential actors within the network.

- *Network integration*: Fusing various networks or systems into a cohesive network is referred to as network integration. Network integration aims to streamline network management, lower expenses, and increase network infrastructure's efficiency, dependability, and scalability. Local area networks (LANs), wide area networks (WANs), and wireless networks are just a few examples of the various network types that can be integrated. It may also entail the fusion of many network technologies, including cellular, Wi-Fi, and Ethernet. Network planning, design, implementation, and testing are often included in the network integration process (Zhang et al., 2020). The network integration team will assess the current networks and create a strategy for integrating them into one network throughout the planning and design phase. This could entail building new network infrastructure, configuring network devices, and upgrading hardware and software (Jabeen et al., 2016; Zhang et al., 2020). The network integration team will implement the modifications after creating the plan and testing the new network to ensure it operates correctly. This could entail evaluating the network's

functionality, security, and dependability in various scenarios. Managing numerous networks or network systems requires enterprises to implement network integration as a necessary procedure. Organizations can enhance network performance, lower expenses, and simplify network management by combining various networks into a single, coherent network(Jabeen et al., 2016; Zhang et al., 2020).
- *Boundary spanners:* Boundary spanners are people or organizations that work at the intersection of various organizational divisions or between an organization and its external environments, such as clients, vendors, or other stakeholders. Their job is to close the gaps between various viewpoints and interests while facilitating communication and collaboration across many domains. Customer service agents, salespeople, public relations experts, and interdepartmental coordinators are a few examples of boundary spanners(Jabeen et al., 2016; Zhang et al., 2020).

3. Background of social network analysis

The easiest way to understand social networks is to think about your social networks. In the current generation, everyone has a social network; people connect. This includes networks related to friends, networks of peers, and networks of people you know in various organizations. It is normal to become members of different social networks. So, a network of friendship becomes a relationship, a network of advisors becomes another, and a network of enmity becomes another. Segmenting social networks by association helps researchers in the way they ask respondents questions. For example, start asking a series of specific questions about the different types of social networks you have. Respondents will easily visualize your other social networks (Panaite et al., 2022; Jabeen et al., 2016; Yildirim & Esen, 2023; Zhang et al., 2020). In most cases, these questions generate a new list of names, and sometimes the same person appears more than once in an answer.

Social network analysis examines relationship models (network structure) and interactions between them at the macro level while studying human behavior at the micro level. Singular conduct is a result of the social network. As people form, maintain, and end relationships, social networks shape the overall network structure while limiting individual freedom of choice (Yildirim & Esen, 2023). Depending on how valuable the relationship under study is as an instrument, different network configurations and locations will produce various opportunities or, conversely, additional limitations. Social capital is the base of events created by social relationships (Yildirim & Esen, 2023).

Several measures have been developed in SNA to characterize and compare network structure and position. Depending on what factors define differences in opportunity structure, investigations can focus on differences in centrality, studies of strongly connected clusters, structurally equivalent network positions, or unique positions (Stockman et al., 2013). For example, a network structure can be compared with different measures to examine its efficiency in achieving a goal. Statistical networking models can also be used to test systems against empty models, estimate parameters and, more recently, test the network effects of different stimulus structures. The study of network exchange theory focuses on how network structure impacts exchange rates and the decision to choose one exchange over another. The significant consequence is connected explicitly to how likely social actors are to exclude others (Stockman et al., 2013). The effects of network architectures are instance-dependent and cannot be generalized without considering the situation, as shown by switching back to network theory. Fusing exchange theory with social networks has also achieved great success in political networks. According to Laumann, most models are based on the Coleman exchange model and limit the influence flow to network links. These models may define the stages from micro-actions to macro-effects, forecast decision results, identify the social users' strengths, and assess the importance of decisions. Many lingering theoretical issues are attempted to be addressed by more current models (Yildirim & Esen, 2023).

A broader theory of negotiation offered by Stockman et al. (2013) claims that the three basic negotiating techniques—persuasion, social network manipulation, and coercion—are linked to three different types of social networks:

1. *Information power networks:* In SNA, power networks refer to how power and influence are distributed among individuals and groups within a social network. Information power networks focus on how information control and dissemination contribute to power dynamics within a network. Information power networks can be analyzed by examining the flow of information between nodes (individuals or groups) within a network. This can include identifying key players who have access to and control critical information and analyzing patterns of communication and information sharing within the network. In information power networks, the ability to prevent or restrict access to information can be a significant source of power. For example, individuals with access to

sensitive information may use the information to influence others or gain influence in negotiations.

Conversely, people connected to important information may be disadvantaged online (Stockman et al., 2013). SNA techniques can be used to identify patterns in information flow and the distribution of power in networks and can be helpful in various contexts, including business, politics, and social movements (Tuffery, 2023). By understanding the energy information networks in each social context, individuals and organizations can better position themselves for success and manage power dynamics more effectively (Tuffery, 2023).

1. *Exchange power networks:* Exchange power networks are social networks in which power is distributed through exchanging resources, such as goods, services, or information. In exchange power networks, individuals or groups have power and influence based on their ability to provide or control access to resources valued by others. SNA techniques can analyze exchange power networks. One way to do this is by examining the patterns of resource exchange and the relationships between individuals or groups who exchange resources. For example, individuals who can provide a valuable resource, such as information or access to a key decision-maker, may be seen as more potent within the network. Exchange power networks can also be analyzed by examining the types of resources exchanged and the regulations and rules governing the exchange process (Singh et al., 2023/2023). For example, in some networks, reciprocity may be a fundamental norm, meaning that individuals are expected to give back roughly what they receive. In other networks, the standards and rules around exchange may be more complex, with individuals using their power to negotiate more favorable terms or to create asymmetrical exchange relationships. Exchange power networks exist in various contexts, including economic markets, social movements, and political systems (Dwivedi et al., 2022; Gupta et al., n.d.; Singh et al., 2023/2023). By understanding the exchange of power networks within a given context, individuals and organizations can better position themselves to succeed within the network and negotiate power dynamics more effectively (Dwivedi et al., 2022; Gupta et al., n.d.; Singh et al., 2023/2023).
2. *Hierarchical power networks:* Hierarchical power networks refer to social networks in which power and influence are distributed based on the hierarchical position or rank of individuals or groups within a formal organization or system. Hierarchical power networks are characterized by clear lines of authority and standard rules and procedures for

decision-making and resource allocation. In hierarchical power networks, power flows down from top to bottom. Power is often concentrated at the top of the hierarchy, with those in leadership positions having the most significant degree of power and control (Ryan, 2023; Tuffery, 2023). SNA can analyze hierarchical power networks by examining the hierarchy structure and the relationships between individuals or groups at different levels. For example, SNA can identify critical players or decision-makers in a network and the extent to which power is centralized or layered (Kozitsin, 2023). Hierarchical power networks exist in many contexts, including governments, corporations, and nonprofit organizations. By understanding hierarchical networks of power in each context, individuals and organizations are better positioned to succeed in the network and manage power dynamics more effectively. However, it is also essential to recognize that hierarchical power networks can create power imbalances and marginalize individuals or groups lower in the hierarchy (Kozitsin, 2023; Ryan, 2023; Tuffery, 2023).

4. Depicting SNA via graphs and graph theory

Graphs, specifically graph theory or the graph of words, are powerful tools for understanding and visualizing networks that can be applied to analyze social networks. We can use many processing forms, but one such can be the natural processing language which we can represent in the graph. We use vocabulary and ideas from graph theory, a branch of mathematics that quantifies networks when we graph social networks and represent social networks as graphs (Bhardwaj et al., 2023; Garg, 2022; Garg & Kumar, 2018; Jancey et al., 2023; Kozitsin, 2023; Ryan, 2023; Tuffery, 2023; Verd, 2023). Graph theory is not the same as SNA, but many underlying ideas and terms have been taken from it. Therefore, it is worthwhile to spend some time getting to know the fundamentals of the subject(Kozitsin, 2023). SNA is a division of network science that seeks to understand complex social networks' structure, dynamics, and evolution. SNA can represent social networks clearly and concisely using graph theory. Graph analytic techniques used in SNASNA are typically focused on studying the interconnectivity, distributions, and clustering patterns of a network within itself (Kozitsin, 2023). A graph consists of nodes and edges and can be used to represent a variety of networks. We can connect nodes to create graphs that show how the social network works. This can be used to uncover meaningful relationships between people and discover influential people and groups on your web (Kozitsin, 2023; Ryan, 2023).

5. Benefits of graphical algorithms

- Graph algorithms can be used to analyze the network structure further. These algorithms can help highlight the most influential individuals in a networking system by determining relationships between individual nodes and measuring centrality (Yadav et al., 2015).
- Graph algorithms can also help predict a network's evolution (Kong et al., 2023).
- Graphs are also helpful for visualizing social networks, helping to understand better the network's structure and dynamics (Yadav et al., 2015).
- Graphs can represent relationships between people, such as social networks or more complex networks like the Internet. Graphing these networks provides a better understanding of their dynamics and how they interact (Kong et al., 2023; Yadav et al., 2015).
- In conclusion, graph theory is a powerful tool for understanding and visualizing networks and can be applied to various networks, including social networks. Graph algorithms can analyze network structure, identify essential relationships, and predict how networks evolve (Kong et al., 2023; Singh et al., 2023/2023; Yadav et al., 2015).

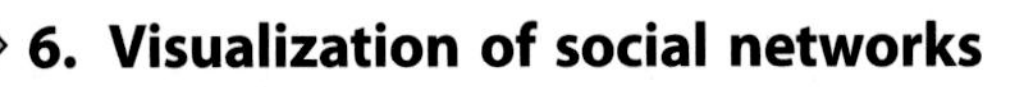

6. Visualization of social networks

Visualization of social networks allows us to understand their structure and dynamics better.

1. *NETWORK MATRICES:* One of the most generalized or used ways to represent a graph is via an adjacency matrix for anything related to social networking. This involves using a square matrix where the rows and columns represent the graph nodes, and the cells indicate the presence or absence of edges between node pairs. In an adjacency matrix, if there is a connection between node i and node j, the corresponding cells (i,j) and (j, i) will be assigned a value of 1 (2023/2023b). Conversely, if there is no link between them, the corresponding cells will be given a value of 0. This matrix can then be utilized to compute various graph metrics, such as degree centrality, betweenness centrality, and closeness centrality, as well as to visualize the graph using network visualization

tools. Although SNA techniques have unique characteristics, they share similarities with other types of networks.

2. *ADJACENCY MATRIX:* This matrix is typically a squared matrix used to represent a graph, where the rows and columns correspond to the graph's vertices, and the entry in the (i,j) position means whether there is an edge between vertex i and vertex j. Typically, the entrance is one if there is an edge and 0 if there is no edge, but other values can be used to represent different types of edges or edge weights.
3. *SYMMETRIC ADJACENCY MATRIX:* A symmetric matrix is a type of matrix under the adjacency, which is used to detect undirected graphs, where the matrix is symmetric, meaning the (i, j) entry is equal to the (j, i) access for all i and j. This is because an undirected graph's edge between vertex i and vertex j is equivalent to a boundary between vertex j and vertex i. It remains symmetric for an undirected graph.
4. *ASYMMETRIC ADJACENCY MATRIX:* An asymmetric matrix is a type of adjacency matrix for a directed graph. In this case, the matrix is asymmetric. Asymmetric adjacency matrices are commonly used in directed graphs, where edges have a direction and may not be reversible. (2023/2023a). In an asymmetric adjacency matrix, the (i,j) entry represents the edge from one vertex to another, which may or may not be the same as the edge from vertex j to vertex i.
5. *WEIGHTED MATRIX*: This is the most valuable quantitative method for evaluating a set of options against many rules. It is beneficial when selecting the best choice and considering numerous criteria. In the context of graphs, a weighted matrix is often used to represent the weights or distances associated with the edges between the vertices of a graph. For example, in a weighted adjacency matrix, the entry in the (i,j) position represents the weight or distance associated with the edge between vertex i and vertex j. A weighted decision matrix provides a structured approach to decision-making, which helps to remove subjectivity, emotion, and uncertainty from the process. This allows for more objective and rational decision-making. The matrix enables a more consistent evaluation of alternatives by assigning weights to various criteria.

   Furthermore, the decision matrix allows for easy referencing of decisions in discussions, meetings, or presentations, providing a way to justify the decisions made. This approach may also be known as a weighted scoring model or prioritization matrix. Weighted matrices can be utilized in other applications where values must be assigned to pair elements, such as machine learning, network analysis, and optimization problems.

6. *INCIDENCE MATRIX:* An incidence matrix represents a graph using edges and vertices. The matrix columns correspond to the vertices, while the rows correspond to the edges. If an edge "I" is incident on a vertex j, the entry in the (i,j) position is 1. If the edge "I" is incident on vertex j, but in the opposite direction, the entry is −1. If the edge "I" is not incident on vertex j, the access is 0. Incidence matrices are commonly used in graph theory to study graph properties such as connectivity, planarity, and graph coloring. They have applications in many fields, including computer science, physics, and engineering.

7. Centrality and composition

- *Quantifying social structure*: Social structure can be quantified through various methods, including network analysis and statistical modeling. One approach is to use network analysis to study the patterns of social relationships within a group or society. This involves identifying nodes (individuals or groups) and edges (relationships or connections) within the network and analyzing their structural properties, such as centrality, density, and clustering (2023/2023a). Other methods for quantifying social structure include regression analysis, factor analysis, and multilevel modeling, which can identify patterns and relationships among social variables such as income, education, occupation, and social status. The approach and methods will depend on the research question and available data.
- *Degree centrality*: SNA uses the concept of "degree" to refer to a node's level of connectivity in a network. The degree of a node can be either incoming, where all nodes are interconnected, or the number of connected nodes, depending on the type of network being analyzed. This measure helps identify influential nodes or network "influencers" based on their outreach or centrality.
- *Betweenness centrality*: A degree of power in SNA refers to the amount or bunch of connections the vertex or the node has in a network. The degree of a node in a network can be used to measure its centrality or outreach. This measure is commonly used to identify virtual nodes or network influencers. Depending on the type of network being analyzed, the degree of nodes can be either the number of connected nodes or the incoming degree; otherwise, the number of connected nodes or the outgoing degree.
- *K-path centrality*: The metric described is k-path centrality, which measures the number of k-length paths that pass through a node in a

network. This measure is used to identify critical nodes for maintaining connectivity in a network and can be particularly useful in identifying essential nodes of transportation, communication, or social networks. A k-path is a sequence of k nodes with no repeats in which the first node is connected to the second, the second to the third, and so on. The number of k-length paths that pass through a node, weighted by their frequency or importance, is called its K-path centrality. Nodes with the most remarkable centrality are connected by multiple k-length paths in the network, indicating their importance in maintaining its overall structure and connectivity (SNA Applications, 2023). This centrality method is a more common measure of centrality than other measures, such as degree or betweenness, because it considers the importance of longer paths beyond the immediate neighbors of a node.

- *CALCULATION*: The calculation of K-path centrality involves counting the number of k-length paths that pass through each node in the network. For a given value of k, the K-Path of any node is defined as K-Path centrality(node) = sum of the number of k-length paths that pass through the node. Calculating K-path centrality depends on the definition of "importance" or "weight" assigned to each course. For example, the weight of each k-length path could be based on its frequency, distance, or some other measure of importance. One common approach is to use the inverse of the path length as the weight so that longer paths are given less weight than shorter paths. Once the weights have been assigned, the K-path centrality of each node can be computed by the summations of the weights of all k-length routes that pass through it. Computing k-path centrality can be computationally expensive for large networks because it involves counting the number of paths with a length "k" for every pair of nodes. Therefore, efficient algorithms and approximations have been developed to estimate the K-path centrality of large networks (Kong et al., 2023; Singh et al., 2023/2023; Yadav et al., 2015).

8. The foundations of network connectivity

Network connectivity is based on several core technologies and protocols: network architecture. The wireless network connections of these networks are different because of the large amount of data with the short text [46]. The design and layout of a network, including its physical and logical structure, determine how devices connect and communicate.

- Transmission medium: A physical medium that transmits data, such as copper wires, fiber optic cables, and radio signals.
- Network protocol: Rules and procedures governing transmission between two or more devices on a network. Examples include TCP/IP, HTTP, FTP, and DNS (Nabiilah et al., 2023).
- Network devices: The hardware and software components used to create and manage networks, including routers, switches, firewalls, and network operating systems.
- Addressing and naming: A system for identifying and naming devices and resources on a network, including IP addresses, domain names, and MAC addresses (Shah et al., 2023).
- Network security: Measures to safeguard networks from unauthorized access and other security breaching threats, including firewalls, intrusion detection systems, and encryption. Network security is typically performed using a combination of hardware and software-based tools and techniques. Some of the essential methods used for network security include:
- Firewalls: A firewall is a hardware or software-based solution that guards and filters incoming and outgoing network traffic based on predefined security terms and conditions (Gupta & Gupta, 2023/2023).
- Intrusion detection/prevention systems (IDS/IPS): This method checks network traffic for signs of potential security threats and can alert administrators or take automated actions to prevent attacks.
- Virtual private networks (VPNs): A VPN is a secure connection between two or more devices. It allows for secure and encrypted communication between devices on a network (Gupta & Gupta, 2023/2023; Nabiilah et al., 2023; Shah et al., 2023).
- Access control: Access control measures limit permission to network resources typically dependent on user credentials or other security factors such as device or location.
- Encryption: Encryption converts data into a coded format to prevent unauthorized access. It is commonly used to secure data during transmission over a network.
- Adequate network security involves implementing multiple layers of protection and constantly monitoring and updating security measures to address emerging threats. Network Monitoring: Network monitoring tools monitor network traffic, detect anomalies, and identify potential security threats. These primitives allow you to create, manage, and secure a modern network that enables communication and data exchange

between devices and users worldwide (Bhardwaj et al., 2023; Jancey et al., 2023; Verd, 2023). Regular security audits and training are also required to ensure all users understand and adhere to security best practices (Garg & Kumar, 2018). We can find more details by following (Garg et al., 2022; Hazarika et al., 2023; Kumar et al., 2022; Malviya et al., 2023).

9. Conclusion

In conclusion, networking analysis is a powerful mechanism for understanding and mapping relationships between individuals, organizations, and other entities. It is used to uncover social network structure, measure network members' influence, and predict future trends. SNA has many benefits. It can help identify central nodes in a network, measure the strength of relationships between members, detect communities and cliques, and detect network changes over time. It can monitor online conversations, measure influence and influence dynamics, and see influencers.

On the other hand, SNA also has some drawbacks. It can be time-consuming and require specialized skills to analyze an extensive network. There can also be privacy issues when collecting data for a network. Additionally, it can be challenging to determine data accuracy, and results may be difficult to interpret. In conclusion, SNA can provide insight into social networks by helping identify patterns, relationships, and dynamics. However, it is crucial to consider the costs and challenges associated with SNA when deciding whether to use it. Research on social networking has led to several conclusions. Social networking has a significant impact on the way people communicate, connect, and share information. It positively and negatively affects society, including social behavior, mental health, politics, and business. These networking platforms have raised concerns about privacy, online security, and the spread of misinformation. Finally, social networking will continue to evolve and impact society in new and unforeseen ways. Ongoing research is needed to understand these impacts better and develop strategies to address the challenges posed by social networking.

References

AlKheder, S., Alzarari, A., & AlSaleh, H. (2023). Urban construction-based social risks assessment in hot arid countries with social network analysis. *Habitat International, 131*, Article 102730.

Bhardwaj, A., Gupta, U., Budhiraja, I., & Chaudhary, R. (January 2023). Container-based migration technique for fog computing architecture. In *2023 international conference for advancement in technology (ICONAT)* (pp. 1–6). IEEE.

Bober, D., & Kaproń, H. (2010). New model of electric energy consumption control - its possibilities and results of simulation research. In *Imcic 2010 - international multi-conference on complexity, informatics and cybernetics, proceedings* (Vol. 2, pp. 125–130). International Institute of Informatics and Systemics, IIIS.

Buhai, I. S., & van der Leij, M. J. (2023). A social network analysis of occupational segregation. *Journal of Economic Dynamics and Control, 147*. https://doi.org/10.1016/j.jedc.2022.104593

Centrality measure. (2023). https://ddu.ext.unb.ca/6634/Lecture_notes/Lecture_4_centrality_measure.pdf.

Cook, Karen S., Emerson, Richard M., Gillmore, Mary R., & Yamagishi, Toshio (1983). The distribution of power in exchange networks: Theory and experimental results. *American Journal of Sociology, 89*(2), 275–305. https://doi.org/10.1086/227866

Creating dynamic network views. (2023). https://www.ibm.com/docs/en/network manager/4.2.0?topic=views-creating-dynamic-network.

Data collection guide. (2023). https://socilyzer.com/guide/data-collection.

Data collection methods. (2023). https://www.simplilearn.com/data-collection-methods-article.

Dwivedi, S. P., Srivastava, V., & Gupta, U. (2022). Graph similarity using tree edit distance. In *Lecture notes in computer science (including subseries lecture notes in artificial intelligence and lecture notes in bioinformatics)* (Vol 13813, pp. 233–241). Springer Science and Business Media Deutschland GmbH. https://doi.org/10.1007/978-3-031-23028-8_24

Electronic communication network. (2023). https://www.Investopedia.com/terms/e/ecn.asp.

Encryption algorithm. (2023). https://www.encryptionconsulting.com/encryption-and-decryption/.

Garg, M. (2022). *An event detection technique using social media data.* arXiv. https://doi.org/10.48550/arXiv.2208.13101

Garg, M., Gupta, A. K., & Prasad, R. (Eds.). (2022). *Graph learning and network science for natural language processing.* CRC Press.

Garg, M., & Kumar, M. (2018). The structure of word co-occurrence networks for microblogs. *Physica A: Statistical Mechanics and Its Applications, 512*, 698–720.

Graph theory and applications. (2023). https://towardsdatascience.com/social-network-analysis-from-theory-to-applications-with-python-d12e9a34c2c7.

Gupta, U., & Gupta, D. (2023). An efficient implicit Lagrangian twin bounded support vector machine. In *International journal of advanced intelligence paradigms.* Original work published 2023.

Gupta, U., Pantola, D., Bhardwaj, A., & Singh, S. P. (n.d.). Next-generation networks enabled technologies. Next generation communication networks for industrial internet of things systems.

Hazarika, B. B., Gupta, D., & Gupta, U. (May 2023). Intuitionistic fuzzy kernel random vector functional link classifier. In *Machine intelligence techniques for data analysis and signal processing: Proceedings of the 4th international conference MISP 2022* (Vol 1, pp. 881–889). Singapore: Springer Nature Singapore.

Incidence matrix. (2023). https://bookdown.org/omarlizardo/_main/3-6-the-incidence-matrix.html.

Intro to SNA. (2023). http://www.orgnet.com/sna.html.

Jabeen, F., Khusro, S., Majid, A., & Rauf, A. (2016). Semantics discovery in social tagging systems: A review. *Multimedia Tools and Applications, 75*, 573–605.

Jancey, J., Vidler, A. C., Leavy, J. E., Chamberlain, D., Riley, T., Pollard, C. M., ... Blackford, K. (2023). Understanding prevention networks in a local government area: Insights from a social network analysis among western Australian nutrition, physical activity, and obesity prevention programs. *Health Promotion Practice, 24*(1), 103–110.

Knoke, David, & Yang, Song (2020). *Social network analysis*. SAGE Publications. Original work published 2020.

Kong, Xiangjie, Shi, Yajie, Yu, Shuo, Liu, Jiaying, & Xia, Feng (2019). Academic social networks: Modeling, analysis, mining and applications. *Journal of Network and Computer Applications, 132*, 86–103. https://doi.org/10.1016/j.jnca.2019.01.029

Kong, Y. X., Wu, R. J., Zhang, Y. C., & Shi, G. Y. (2023). Utilizing statistical physics and machine learning to discover collective behavior on temporal social networks. *Information Processing & Management, 60*(2). https://doi.org/10.1016/j.ipm.2022.103190

Kozitsin, I. V. (2023). Opinion dynamics of online social network users: A micro-level analysis. *Journal of Mathematical Sociology, 47*(1), 1–41.

Kumar, S., Gupta, S., & Gupta, U. (December 2022). Discrete cosine transform features matching-based forgery mask detection for copy-move forged images. In *2022 2nd international conference on innovative sustainable computational technologies (CISCT)* (pp. 1–4). IEEE.

Majeed, Abdul, & Rauf, Ibtisam (2020). Graph theory: A comprehensive survey about graph theory applications in computer science and social networks. *Inventions, 5*(1). https://doi.org/10.3390/inventions5010010

Malviya, L., Mal, S., Kumar, R., Roy, B., Gupta, U., Pantola, D., & Gupta, M. (2023). Mental stress level detection using LSTM for WESAD dataset. In *Proceedings of data analytics and management: Icdam 2022* (pp. 243–250). Singapore: Springer Nature Singapore.

Nabiilah, Ghinaa Zain, Prasetyo, Simeon Yuda, Izdihar, Zahra Nabila, & Girsang, Abba Suganda (2023). BERT base model for toxic comment analysis on Indonesian social media. *Procedia Computer Science, 216*, 714–721. https://doi.org/10.1016/j.procs.2022.12.188

Network traffic analyser. (2023). https://www.manageengine.com/products/netflow/free-network-traffic-analyzer.html.

Nooraie, R. Y., Mohile, S. G., Yilmaz, S., Bauer, J., & Epstein, R. M. (2021). Social networks of older patients with advanced cancer: Potential contributions of an integrated mixed methods network analysis. *Journal of Geriatric Oncology, 12*(5), 855–859.

Panaite, V., Yoon, S., Devendorf, A. R., Kashdan, T. B., Goodman, F. R., & Rottenberg, J. (2022). Do positive events and emotions offset the difficulties of stressful life events? A daily diary investigation of depressed adults. *Personality and Individual Differences, 186*, Article 111379.

Prell, Christina (2012). *Social network analysis: SAGE publications* (Vol 9781412947152). Original work published 2012.

Reachability matrix. (2023). https://www.sciencedirect.com/topics/computer-science/reachability-matrix.

Ryan, L. (2023). Conceptualising migrant networks: Advancing the field of qualitative social network analysis. In *Social networks and migration* (pp. 14–36). Bristol University Press.

Shah, Zohal, Chen, Chen, Sonnert, Gerhard, & Sadler, Philip M. (2023). The influences of computer gameplay and social media use on computer science identity and computer science career interests. *Telematics and Informatics Reports, 9*. https://doi.org/10.1016/j.teler.2022.100040

Singh, S., Bhardwaj, A., Budhiraja, Gupta, U., & Gupta, I. (2023). Cloud-based architecture for effective surveillance and diagnosis of COVID-19. In *Convergence of cloud with AI for big data analytics*. https://doi.org/10.1002/9781119905233.ch4. Original work published 2023.

SNA applications. (2023). https://towardsdatascience.com/social-network analysis-from-theory-to-applications-with-python-d12e9a34c2c7.
SNA and its use case. (2023). https://www.latentview.com/blog/a-guide-to-social-network-analysis-and-its-use-cases/.
SNA methods and techniques. (2023). https://www.sciencedirect.com/topics/social-sciences/social-network-analysis.
SNA use cases. (2023). https://www.latentview.com/blog/a-guide-to-social-network-analysis-and-its-use-cases/.
Social network analysis. (2023). https://en.wikipedia.org/wiki/Social_ network_ analysis.
Stockmann, D. (2013). *Media commercialization and authoritarian rule in China*. Cambridge University Press.
TCP/IP model. (2023). https://www.guru99.com/tcp-ip-model.html.
TCP/IP. (2023). https://www.techtarget.com/searchnetworking/definition/TCP-IP.
Toomet, O., Van Der Leij, M., & Rolfe, M. (2013). Social networks and labor market inequality between ethnicities and races. *Network Science, 1*(3), 321–352.
Tuffery, S. (2023). *Social network analysis*.
Valeri, M., & Baggio, R. (2021a). Italian tourism intermediaries: A social network analysis exploration. *Current Issues in Tourism, 24*(9), 1270–1283.
Valeri, M., & Baggio, R. (2021b). Social network analysis: Organizational implications in tourism management. *International Journal of Organizational Analysis, 29*(2), 342–353. https://doi.org/10.1108/IJOA-12-2019-1971
Verd, J. M. (2023). Using a hybrid data collection tool: Analysis of youth labour market trajectories integrating quantitative, qualitative and social network data. *International Journal of Social Welfare, 32*(1), 9–19.
VPN and firewall. (2023). https://www.geeksforgeeks.org/relationship-between-vpn-and-firewall/.
Wang, Heli, Tong, Li, Takeuchi, Riki, & George, Gerard (2016). Corporate social responsibility: An overview and new research directions. *Academy of Management Journal, 59*(2), 534–544. https://doi.org/10.5465/amj.2016.5001
Wasserman, Stanley, & Faust, Katherine (1999). *Social network analysis methods and applications: Syndicate of the university of cambridge*. Original work published 1999.
Webber, Q. M., Dantzer, B., Lane, J. E., Boutin, S., & McAdam, A. G. (2023). Density-dependent plasticity in territoriality revealed using social network analysis. *Journal of Animal Ecology, 92*(1), 207–221.
Weighted decision matrix. (2023). (Original work published 2023).
What is data collection. (2023). https://www.simplilearn.com/what-is-data-collection-article.
Yang, S., & Knoke, D. (2001). Optimal connections: Strength and distance in valued graphs. *Social Networks, 23*(4), 285–295.
Yadav, S., Mishra, R., & Gupta, U. (2015). Performance evaluation of different versions of 2D Torus network. In *Conference proceeding - 2015 international conference on advances in computer engineering and applications, ICACEA 2015* (pp. 178–182). Institute of Electrical and Electronics Engineers Inc. https://doi.org/10.1109/ICACEA.2015.7164691
Yildirim, Ş. S., & Esen, Ü. B. (2023). Social entrepreneurship activities in the tourism sector: Bibliometric analysis and social network analysis. In *Research anthology on approaches to social and sustainable entrepreneurship* (pp. 1253–1274). IGI Global.
Yoon, Sunkyung, Kleinman, Mary, Mertz, Jessica, & Brannick, Michael (2019). Is social network site usage related to depression? A meta-analysis of facebook–depression relations. *Journal of Affective Disorders, 248*, 65–72. https://doi.org/10.1016/j.jad.2019.01.026

Yousefi Nooraie, R., Sale, J. E. M., Marin, A., & Ross, L. E. (2020). Social network analysis: An example of fusion between quantitative and qualitative methods. *Journal of Mixed Methods Research, 14*(1), 110–124. https://doi.org/10.1177/1558689818804060

Zhang, Z., Gao, Y., & Li, Z. (2020). Consensus reaching for social network group decision making by considering leadership and bounded confidence. *Knowledge-Based Systems, 204*, Article 106240.

CHAPTER TWELVE

Human AI: Explainable and responsible models in computer vision

Kukatlapalli Pradeep Kumar, Michael Moses Thiruthuvanathan, Swathikiran K.K. and Duvvi Roopesh Chandra
Department of Computer Science Engineering, School of Engineering and Technology, CHRIST University, Kengeri Campus, Bangalore, Karnataka, India

1. Introduction

The progress of AI is opening up new opportunities in a variety of fields, including corporate business, medical, and education management, among others, to improve people's lives all over the world. Furthermore, it raises further questions about the most effective ways to include fairness, interpretability, privacy, and security into these systems. There are certain contributions for AI engineers to ensure manifestation in quality of design, metric ascertaining, data examination, understanding limitations with respect to model and data, integrated testing and unit testing, and continuous monitoring when considering responsible practices such as fairness, interpretability, and security. Since these issues are far from being resolved, particularly in the realm of AI, they must be reinforced in every AI initiative, interpretation, and unique design (Lavin et al., 2021).

Deep learning—based AI systems are very complicated and hard for many people to understand, but the ideas behind them are very interesting. When it comes to design, very few designs are clear and easy to understand. The science behind each network must be both complicated and easy to understand. Researchers can look at the overall performance of the system and the short- and long-term health of the product by looking at the false positive- and false-negative rates for each subgroup. Deep learning models will look like the data they are trained on. Because of this, it is important for users to study the raw data carefully to make sure they have a full understanding of the input data. The sensitive raw data computing aggregated summaries and anonymous summaries will make sure that the models are compatible while

Emotional AI and Human-AI Interactions in Social Networking
ISBN: 978-0-443-19096-4
https://doi.org/10.1016/B978-0-443-19096-4.00006-7

still protecting the privacy of information like personal, medical, or insurance information.

Deep learning models are often fixed in size and designed to handle certain issues. However, a lesser-known technique of mitigating network alterations will cause the network to be less productive by causing underfitting, overfitting, or dimensionality mismatch difficulties. Integrating the best testing methodologies used in software engineering and quality engineering to guarantee that the AI system works as it should and can be trusted. The best practices for enabling responsible AI are to disassemble the system and test the smaller units in isolation, as well as to combine the smaller units and test them throughout the training and testing phases. As a gold standard practice, one should check that the system is behaving as intended. It is critical to build a system with quality controls in place to prevent or rapidly respond to any unexpected failures (Alzubaidi et al., 2021).

Monitoring the system constantly will ensure that the model incorporates real-world performance and accepting user feedback shorten the defects. Any model of the world is nearly by definition flawed. Integrating time into the product road plan for problem resolution by considering both short- and long-term options will enforce problem redressal. Before upgrading a deployed model, assessing the differences between the candidate model and the deployed model needs differentiation and the impact of the update on the overall system quality and user experience.

Notably, although responsible AI in the deep learning workflow is a crucial aspect, designing products with AI ethics in mind requires a mix of technical, product, regulatory, procedural, and cultural considerations. These issues are diverse and sociotechnical at their core. For instance, issues of fairness may often be traced back to biases in the world's fundamental processes. As a result, proactive AI responsibility initiatives need not be just improvements to measurement and modeling but also modifications to policy and architecture in order to ensure transparency, rigorous review procedures, and a variety of decision-makers who can contribute diverse viewpoints.

Emotional references from a face may be integrated into a computationally intelligent system using an automatic expression recognition system. Artificial intelligence (AI) systems can understand human emotions and respond appropriately. Establishing an emotional intelligent computing model is currently being worked on by researchers to aid in the areas of behavioral science, psychology, telecommunications, instructional technology, vehicle security, and human—computer interface. The introduction of emotional analysis for the student fraternity has resulted in increased enthusiasm and belief in improving comprehension and endurance of students' learning curve from a classroom. Cognition is the process of learning via thinking, experiencing, and recognizing. The process of obtaining knowledge through study, teaching, and analysis is known as learning. Cognitive learning is the process of paying attention to what is being taught and expressing interest or disinterest based on what is being presented. Emotions are conventional techniques of expressing a perceived idea through the face in order to persuade others to understand the state of mind. As a result, emotions serve as indications to understand students' mindsets, which can help tutors or instructors create a suitable curriculum for technology pedagogy, resulting in improved outcome-based learning (Alzubaidi et al., 2021). There are two approaches for recording and quantifying participants' emotions in the learning setting. The first method is self-evaluation through individual reports based on perceived facts. Second, by automating a system that can generate reports on the entire class and its degree of interest. Yet, information obtained via self-evaluation procedures might be manipulated and often untrustworthy since the data can be preprepared per specific necessity. Furthermore, because the reports are typically retrieved after the session, the data given through self-appraisal do not need to be thorough and detailed. To address this issue, auto assessment mechanisms were proposed. Based on recordings taken from a smart classroom, computer vision techniques are utilized to analyze students' moods in the classroom. A slew of tests have been carried out in order to evaluate and understand the emotion data collected, analyzing the growth of emotions in classroom videos for both individuals and groups.

We present specific information regarding explainable AI and responsible AI in relation to ontology elements in this chapter. The next sections present concepts related to literature, ontology connections, and relevant outcomes, followed by a conclusion.

2. Background literature

This section covers two majors aspects linked to the title of the chapter regarding responsible AI and explainable AI in terms of literature review.

2.1 Challenges in the field of education for responsible AI

The impact of AI in the education system is a key topic of discussion in this postdigital era. Even if technology has significantly improved educational quality, the deployment of AI in education should be treated much more seriously. Combining AI and education to achieve digital transformation with the needs of both students and teachers will result in a big breakthrough in education. In order for an AI infrastructure to be ethical and trustworthy, the employees and corresponding organizations involved in the application development process must accept accountability while keeping professional ethics and human values in mind. The ethical and reliable use of AI is governed by the ART principles of accountability, responsibility, and transparency. Concerns about data privacy, security, and appropriate use have emerged as current obstacles in the field of AI in education. Any application of AI without first considering the potential consequences is likely to have undesirable outcomes. Students using virtual assistants as a companion should be aware that the agent's empathic version, which may bring up the student's past successes and failures, could have a chilling effect on the student's development as a learner. In another scenario, when parents desire to access students' personal data, even if it helps the student obtain higher grades and test scores and makes them attend classes on a regular basis, teenagers would demand a certain amount of autonomy to accept responsibility. Collaboration across space, time, culture, and context is facilitated by the digital system. Any concern or doubt that a learner may have can be quickly resolved with the assistance of AI. Problem-solving in a creative manner integrating empathy, logic, and unique thinking is a talent that students must learn through the curriculum. The ability to adapt and learn new skills, which allows one to continue evolving, is also essential (Dignum, 2021).

2.2 Digital health aspects of responsible AI

Healthcare is a significant application of responsible AI, as the main focus of responsible AI is on design, development, and ethical AI that is applicable to the healthcare setting. The primary objective of responsible AI is to increase interpretability and explainability of results while decreasing biases and promoting justice. Responsible AI deployment is especially crucial in the healthcare sector, which addresses people's physical, emotional, and social well-being. Individuals' well-being would benefit from such implementation. Yet, AI decisions and actions will have moral repercussions or sometimes violate ethical principles. Understanding the essential components of responsible AI is important in the realm of digital health and telemedicine. Datasets of varying sizes gathered from recognized medical departments across patient health problems provide us with a comprehensive approach to digital data analysis in healthcare. This, in turn, will aid in making responsible judgments for creating AI applications (Trocin et al., 2021).

2.3 Responsible AI in safeguarding and conservation of living beings

AI and machine learning have advanced to the point where they now cover practically every element of human life, including environmental conservation. Algorithms capable of machine learning can estimate the extinction risk of thousands of species, analyze the global footprint of fisheries, and identify animals and people in field-collected wildlife sensor data. These machine learning models may be trained using Google Earth Engine for satellite data, Movebank for animal tracking data, and other global-scale, conservation-relevant datasets. Despite the fact that AI is transforming the area of nature conservation, the opaque nature of some machine learning algorithms raises the possibility of unintended consequences that might have enormous effects on humans and animals. Before AI and machine learning are implemented in this market, several dangers must be addressed. Among these risks are the difficulty in recognizing an algorithm's hidden assumptions and the absence of transparency when an algorithm is asked to make predictions outside the training data's bounds. Similarly, exploring an algorithm for making a certain decision is challenging. Working together, ML researchers and conservationists are producing improved metrics for gauging the value of any specific algorithm for doing actual conservation, which is a significant advantage for this industry. In conjunction with improved assessment measures, ethical supervision would make AI a transparent and reliable technology. Population viability analysis (PVA) is a prominent example of a method used to forecast the possibility of a species becoming extinct in the future. Incorporating responsible AI into the area of nature conservation will improve openness and guarantee that researchers can demonstrate that they have considered both the generalizability and limitations of their technique (Wearn et al., 2019).

2.4 Explainable AI in making understandable AI

Artificial intelligence is a wide area that predicts outcomes in every industry and facet of life. It is employed in numerous fields like portable application development, cyber security infrastructure, natural language processing, financial sector, and industry revolution 4.0 related elements (Ahmed et al., 2022). Yet, only developers are aware of the machine's accuracy and the basis on which it predicts results utilizing AI and ML algorithms. In this regard, the end-user will benefit more from elicitation and suitable description (Gade et al., 2019). Even with deep neural networks, it is hard to produce accurate explanations for judgments made; hence, "Explainable AI" (XAI) should be pursued to have a good effect on communities and companies. XAI focuses on transparency, causality, bias, fairness, and safety and may be used to provide explanations of social significance and rights. The explainable AI component enables one to comprehend the output of a decision-making system (Hagras, 2018).

2.5 Tree-based representations in explainable AI

Not much thought has been given to the rationale behind the predictions made by popular tree-based predictive models such gradient boosted trees, decision trees, and random forests. In order to make tree-based models more easily understood, three different enhancements are made. A game-theoretic approach that can find optimal explanations in polynomial time is an original method of explanation that uses numbers to measure the results of interactions between nearby features. A fresh set of methods for deciphering global models' structure, wherein several regional causes are accounted for in each forecast. Then, these techniques are put to use in an effort to address problems in medical machine learning. We show that by drawing on local explanations; we may present global structure while maintaining local model consistency. The TreeExplainer model takes the black box prediction model as input and outputs the white box explanation using information gleaned from these several local techniques. Improvements in the interpretability of tree-based representations have far-reaching ramifications (Lundberg et al., 2020; Holzinger et al., 2022; Ehsan et al., 2022) due to the prevalence of the aforementioned learning features connected with them. In order to better comprehend how AI should be deployed in the actual world and how to restrict or prevent the potentially damaging impacts of doing so, the authors take a perspective that concentrates on the negative elements of AI. In this editorial, we expand on the idea of responsible AI to discuss the myriad ways in which AI might have unintended repercussions and to propose new directions for future IS research that could help us learn more about how to prevent or lessen the impact of these outcomes. Also, we provide some recommendations for future directions in this line of study (Mikalef et al., 2022).

2.6 Explainable AI for student emotions

There has not been much research on how to effectively recognize and classify students' real-time emotional states in educational settings. It is difficult to categorize the feelings of children in a classroom situation. The ardor of collective delight (relevant placement of persons in the picture) is determined using a weighted group expression model that takes into account both global and local situations (Dhall et al., 2015). Using hand, body, and scene information, this model is improved further for multimodal group effect analysis (Huang et al., 2019). Unfortunately, this study is limited to happy types, and the utilization of local factors in classroom situations is irrelevant affective state detection and classification.

There are various extant works that use text, audio, and video data to recognize students' affective moods. The majority of unobtrusive computer vision-based affective state analysis tools for pupils exclusively employ facial expressions. While there are few works on examining students' affective states using cutting-edge methodologies, there are several cutting-edge techniques on exploring basic emotion recognition in different other domains. Michael et al. proposed a multimodal paradigm for video processing that employs video summarizing and microemotional categorization to reduce computing requirements. Deep networks differ in many ways, including feature extraction, network depth, network categorization, and many others (Thiruthuvanathan & Krishnan, 2022). Ashwin et al. proposed an architecture which utilizes participants' facial gestures, hand and body movements for observing different states such as engaged, boredom, and neutral. The simulation results provide an accuracy close to 86% and 70% for still and dynamic states related to classroom data (Ashwin & Guddeti, 2020).

3. Ontology and explainable AI

Ontologies play a key role in making a concept understood by a wide variety of knowledge workers associated or surrounded by it. An explicit specification of conceptualization is called ontology or simple words it is called "study of existence" (Derave et al., 2022). Explainable AI as a technology is focused on incident response by fragmenting the parts to study about an issue in a black box to white box environment. This has roots in connecting the definition of ontology which specifies study of existence. We bring in a notable bond in this regard with regards to AI which is explainable and ontology. Descriptions, semantics, grammar, and meaning of a concept can be visualized in the form of understandable notations to the people of interest. Responsible AI aims at making systems responsible and resilient which basically avoids a catastrophe where as explainable AI provides an insight into trouble shooting answering the question "what has gone wrong?" The primary approaches, which consist of providing an interpretation of the model's outputs and monitoring the model's decisions to detect and react to unfair behaviors are described in more detail in order to compare our system within state-of-the-art related frameworks. These approaches monitor the model's decisions in order to detect and react to unfair behaviors. It is essential to evaluate our system in light of these other frameworks since doing so enables us to gain a deeper comprehension of the ways in which our system might be enhanced. In conclusion, a fresh new trust and reliance scale is presented for the purpose of evaluating the system, and a usability test is carried out. Both of these are done in order to determine how effective the system is (Purificato et al., 2023) (Fig. 12.1).

The authors in (Dwivedi et al., 2023) examine cutting-edge programming strategies for XAI and describe the steps of XAI development in a normal machine learning development process. We describe the various XAI approaches and, utilizing this taxonomy, discuss the primary distinctions between the available XAI strategies.

To combat security concerns in AI-based systems, new research methodologies must be implemented. This project aims to investigate AI threats that are "intentionally malevolent." It also discusses problem conceptualization

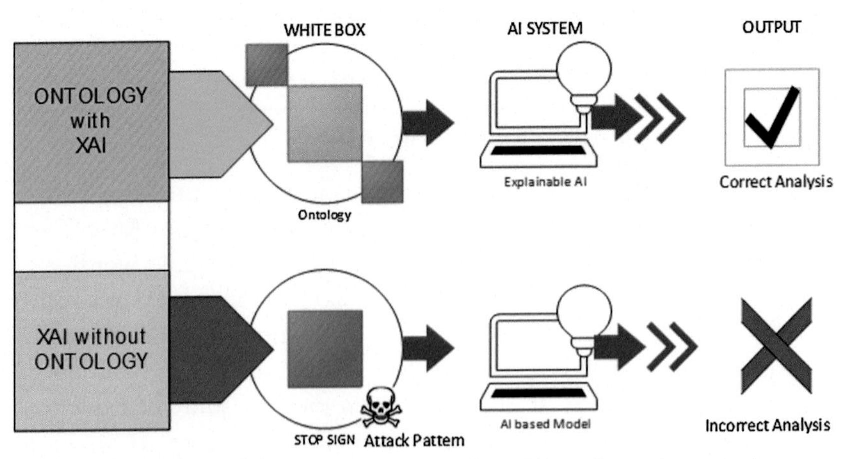

Figure 12.1 Model accuracy representation of three-layered approach.

and solutions for AI attacks using digital forensic techniques (Manasa & Kumar, 2022). The authors focus on interpretability approaches for machine learning; specifically, a literature review and taxonomy of these methods are offered, as well as links to their programming implementations, with the intention that this survey will serve as a resource for both theorists and practitioners (Linardatos et al., 2021).

Ontology incorporated with XAI as one case and XAI system without ontology is depicted in Fig. 12.1. The first case has a better analysis in providing the output structure, whereas the second case has less probability of giving a better understanding of the allocated system. Thus, we invoke ontologies combined or implemented with XAI systems to produce better results in view of computer vision.

4. Results and discussion

In this part, we explain explainable AI in the context of a computer vision simulation experiment involving picture classification. Deep learning is connected with computer vision, which has several applications including autonomous vehicle driving, smartphone photography, security intelligence, etc. In computer vision, image categorization and classification is one of the supervised learning techniques.

4.1 Image categorization using ResNet-50 network

The following is an image classification model utilizing ResNet-50, a 50-layer convolutional neural network (CNN). After loading the pretrained model, three distinct models with 3, 5, and 7 extra layers were generated. By evaluating these models during compilation and training, we have reached a conclusion regarding how the addition of these layers will impact the correctness of the created model. In addition to describing what we are doing and how it is functioning, we also provide the outcomes achieved during the design and evaluation of these models. This facilitates the experiment's explainable and responsible AI field. The dataset we are utilizing contains photos of flowers belonging to five distinct classes, namely, daisy, dandelion, roses, sunflowers, and tulips. The photos of the flowers are separated into five folders that correspond to their various classes.

4.1.1 Three-layer perspective

In this regard, three extra layers were added to the ResNet50 model. It consists of a fully interconnected layer and an output layer that serves as a

Table 12.1 Summary table representation of three-layered approach.

Layer (category)	Output shape	Parameters
resnet50	(Nill, 2048)	23587712
Flatten	(Nill, 2048)	0
Dense	(Nill, 128)	262272
dense_1	(Nill, 5)	645

foundation for actual learning. At the output layer, we employ the softmax activation function, and there are five output neurons corresponding to the five classes in the available data. The values associated with the layer type, output shape, and parameter variables are displayed in Table 12.1.

Model: "sequential".

Total attributes: 23,850,629

Total trainable attributes: 262,917

Total nontrainable attributes: 23,587,712

Once the model becomes complete and executable, it is trained over a period of 10 epochs. The details collected after training it for 10 epochs are displayed in Table 12.2 below. For each epoch, it includes information such as time, train loss, train accuracy, validation accuracy, and validation loss.

The training and validation accuracy in the above table is represented in the form of a graph for better analysis. A model accuracy graph is generated and shown in Fig. 12.2 with epochs on X-axis and accuracy on Y-axis.

Table 12.2 Evaluation table representation of three-layered approach.

Epoch	Time	Loss	Accuracy	val_loss	val_accuracy
1	139s	0.7389	0.7469	0.4433	0.8460
2	132s	0.3152	0.8900	0.3436	0.8638
3	130s	0.2027	0.9329	0.3489	0.8678
4	126s	0.1377	0.9588	0.3706	0.8706
5	129s	0.0898	0.9758	0.3891	0.8624
6	131s	0.0657	0.9830	0.3738	0.8678
7	133s	0.0388	0.9946	0.3961	0.8787
8	133s	0.0232	0.9990	0.3955	0.8815
9	134s	0.0156	0.9997	0.4015	0.8828
10	141s	0.0106	1.0000	0.3880	0.8883

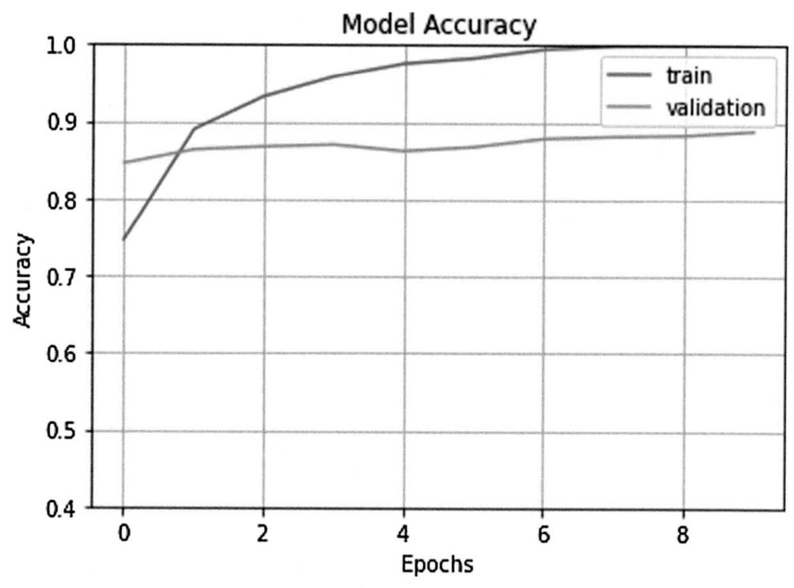

Figure 12.2 Model accuracy representation of three-layered approach.

4.1.2 Five-layer perspective

In this approach, five extra layers were added to the ResNet50 model. Table 12.3 depicts an overall description of the model's considered properties.

Model: "sequential".

Total attributes: 23,883,653

Total trainable attributes: 295,941

Total nontrainable attributes: 23,587,712

Upon adding the additional five layers, the model should be compiled and trained for 10 epochs. The details collected after 10 epochs of training

Table 12.3 Summary table representation of a five-layered approach.

Layer (category)	Output shape	Parameter
resnet50	(Nill, 2048)	23587712
Flatten	(Nill, 2048)	0
Dense	(Nill, 128)	262272
dense_1	(Nill, 128)	16512
dense_2	(Nill, 128)	16512
dense_3	(Nill, 5)	645

Table 12.4 Evaluation table representation of five-layered approach.

Epoch	Time	Loss	Accuracy	val_loss	val_accuracy
1	128s	0.6564	0.7582	0.4282	0.8447
2	143s	0.3224	0.8835	0.3737	0.8597
3	137s	0.2226	0.9240	0.3705	0.8597
4	141s	0.1624	0.9438	0.4439	0.8433
5	137s	0.1023	0.9629	0.4708	0.8719
6	135s	0.0976	0.9639	0.5021	0.8638
7	136s	0.0750	0.9738	0.4752	0.8774
8	136s	0.0313	0.9901	0.5399	0.8747
9	138s	0.0643	0.9792	0.7274	0.8406
10	129s	0.0439	0.9843	0.5664	0.8774

are shown in Table 12.4, which covers information such as time, train loss, train accuracy, validation accuracy, and validation loss for each epoch.

The training and validation accuracy in the above table is represented in the form of a graph for better analysis. A model accuracy graph is generated and shown in Fig. 12.3 with epochs on X-axis and accuracy on Y-axis.

4.1.3 Seven-layer perspective

In this approach, we added seven additional layers to the ResNet50 model. Table 12.5 depicts a seven-layered approach related to the model.

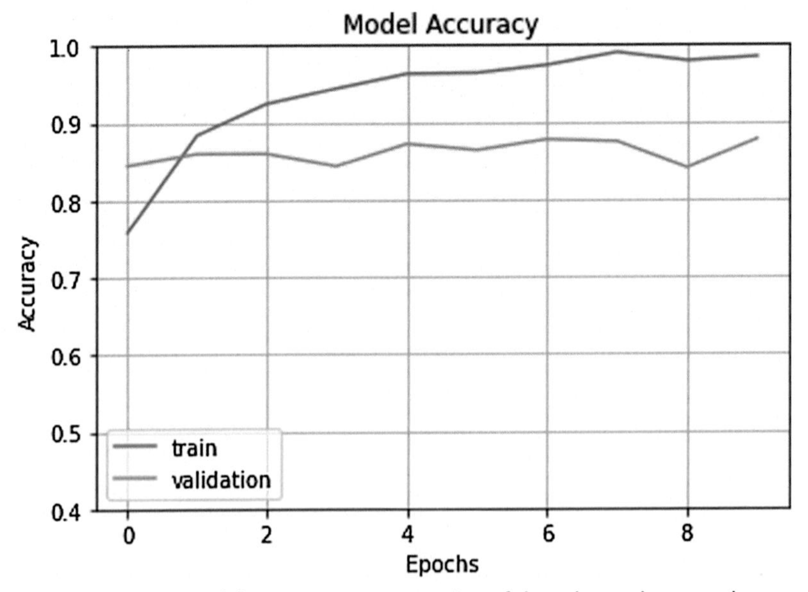

Figure 12.3 Model accuracy representation of three-layered approach.

Table 12.5 Summary table representation of a seven-layered approach.

Layer (category)	Output shape	Parameter
resnet50	(Nill, 2048)	23587712
Flatten	(Nill, 2048)	0
Dense	(Nill, 128)	262272
dense_1	(Nill, 128)	16512
dense_2	(Nill, 128)	16512
dense_3	(Nill, 128)	16512
dense_4	(Nill, 128)	16512
dense_5	(Nill, 5)	645

Table 12.6 Evaluation table representation of three-layered approach.

Epoch	Time	Loss	Accuracy	val_loss	val_accuracy
1	134s	0.6556	0.7585	0.4736	0.8243
2	127s	0.3553	0.8723	0.4060	0.8542
3	127s	0.2526	0.9128	0.3869	0.8501
4	127s	0.1696	0.9401	0.4372	0.8692
5	127s	0.1327	0.9489	0.5312	0.8529
6	127s	0.1290	0.9554	0.4762	0.8651
7	127s	0.1029	0.9601	0.5485	0.8569
8	128s	0.0602	0.9792	0.6066	0.8556
9	127s	0.0384	0.9877	0.5776	0.8733
10	127s	0.0357	0.9857	0.7105	0.8610

Model: "sequential".

Total attributes: 23,916,677

Total trainable attributes: 328,965

Total nontrainable attributes: 23,587,712

On integrating the additional seven layers, the model should be compiled and trained for 10 epochs. The details collected after 10 epochs of training are shown in Table 12.6, which covers information such as time, train loss, train accuracy, validation accuracy, and validation loss for each epoch.

Fig. 12.4 is a visual representation of the model accuracy presented in the tables above, with epochs on the X-axis and accuracy on the Y-axis.

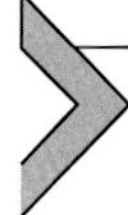

5. Facial emotion detection: application inferring human mental states

The resnet50 that we have built has expanded applications in domains such as social neuroscience, AI for humans, and inferring mental states. This section describes the implementation of this deep learning model for face

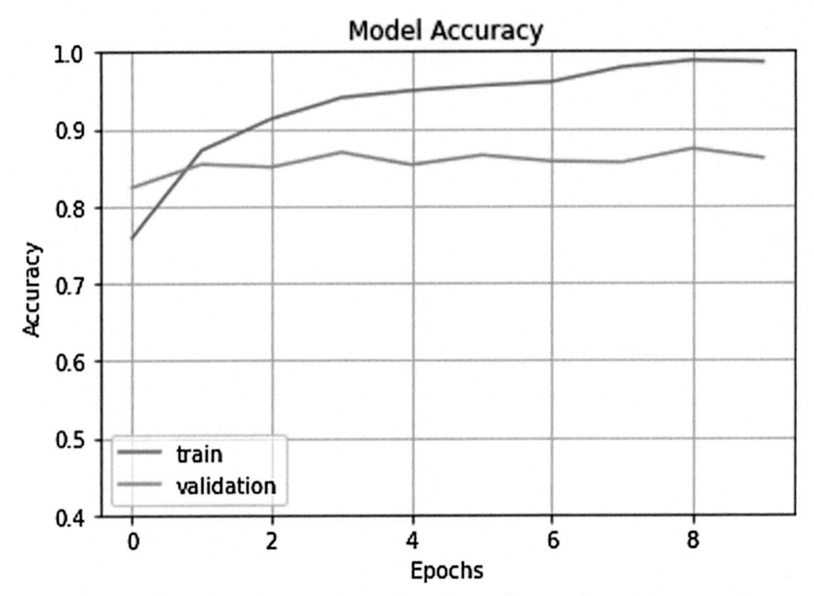

Figure 12.4 Model accuracy representation of seven-layered approach.

emotion recognition. We are focusing on the identification of emotions such as engagement, boredom, confusion, and frustration that a person will experience when present in a classroom similar to the work carried out in (Thiruthuvanathan et al., 2021). So, the training dataset will include these four types of emotion as well as photos of humans exhibiting the corresponding emotion. We created this image classification model using the resnet50's three-layered technique because it yields the best results. Here, we have ran the model for a total of 25 epochs to achieve the highest level of precision (Table 12.7).

Model: "sequential".

Total attributes: 24,638,852

Total trainable attributes: 1,051,140

Total nontrainable attributes: 23,587,712

Table 12.7 Summary information.

Layer (category)	Output shape	Parameter
resnet50	(Nill, 2048)	23587712
Flatten	(Nill, 2048)	0
Dense	(Nill, 512)	1049088
dense_1	(Nill, 4)	2052

Table 12.8 Evaluation information.

Epoch	Time	Loss	Accuracy	val_loss	val_accuracy
1	73s	1.1177	0.6239	0.6612	0.7491
2	66s	0.5092	0.7968	0.4644	0.8255
3	65s	0.4210	0.8369	0.5137	0.7774
4	62s	0.3554	0.8572	0.3965	0.8538
5	63s	0.2790	0.8906	0.3737	0.8491
6	62s	0.2194	0.9130	0.3743	0.8689
7	63s	0.2154	0.9173	0.4387	0.8472
8	62s	0.2205	0.9119	0.4792	0.8425
9	62s	0.2172	0.9166	0.3604	0.8783
10	64s	0.1748	0.9366	0.3445	0.8745
11	65s	0.1674	0.9368	0.3385	0.8887
12	67s	0.1638	0.9375	0.3955	0.8642
13	62s	0.1523	0.9423	0.3770	0.8802
14	65s	0.1368	0.9484	0.3246	0.8896
15	65s	0.1506	0.9420	0.5458	0.8387
16	65s	0.1505	0.9432	0.3584	0.8802
17	65s	0.1227	0.9489	0.3162	0.9000
18	60s	0.1269	0.9489	0.3338	0.8840
19	63s	0.1461	0.9418	0.3115	0.9000
20	62s	0.1044	0.9616	0.3348	0.8821
21	63s	0.0977	0.9618	0.4126	0.8623
22	67s	0.1142	0.9538	0.3149	0.8868
23	66s	0.1036	0.9564	0.3017	0.8887
24	66s	0.0994	0.9566	0.2952	0.9038
25	72s	0.1022	0.9576	0.3870	0.8858

The values that are obtained after training the model for 25 epochs are contained in Table 12.8.

The graph achieved for the train and validation accuracy with respect to epochs is represented below (Fig. 12.5).

A Categorization Report includes evaluation matrix values, which are referred to as f1 score, precision, accuracy, and recall, respectively. These matrix values are provided in the table, along with the macro average and weighted average values for each class. The table also contains all of the classes. This will be of extra assistance in understanding how well the model works on images that come from an unknown dataset.

The multiclass classification technique involves treating the true and predicted classes as multiple variables and determining their correlation coefficients (in a similar way to computing correlation coefficient between any two variables). The stronger the connection between true and anticipated

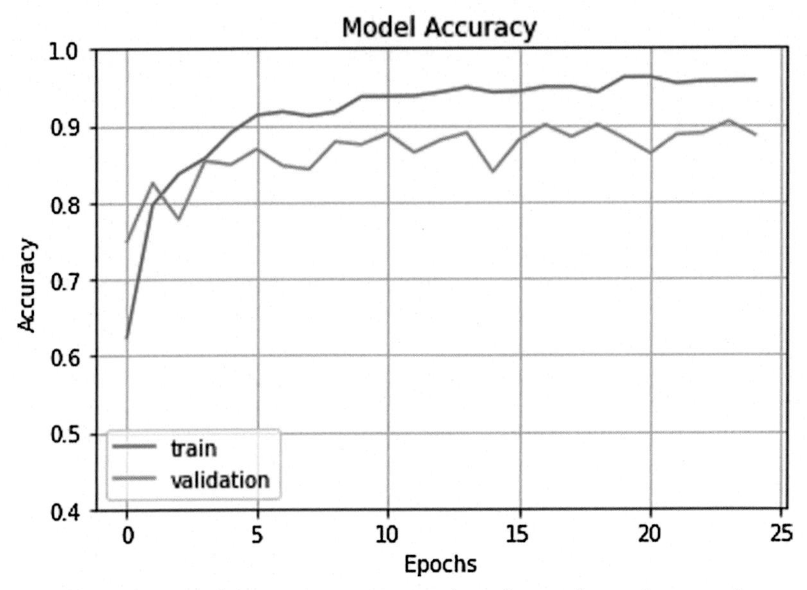

Figure 12.5 Model accuracy representation of seven-layered approach.

values, the better the forecast. This is the √-coefficient often known as the Matthews Correlation Coefficient (MCC) when applied to classifiers MCC.

$$MCC = cs - \frac{cs - \vec{t}\,.\,\vec{p}}{\sqrt{s^2 - \vec{p}\,.\,\vec{p}}\;\sqrt{s^2 - \vec{t}\,.\,\vec{t}}} \tag{12.1}$$

Where, k is the class from 1 to 4, s is the number of samples, c is the number of samples correctly predicted, t_k is the number of times where class k occurred truly, and p_{k} is the number of times class k was predicted. Based on the true-positive, true-negative, false-positive, and false-negative score acquired by the network, the correlation coefficient was recorded at 0.96 (Table 12.9)

Table 12.9 Classification report related values.

	Precision	Recall	F1-score	Support
Boredom	0.92	1.00	0.96	34
Confused	0.96	0.93	0.95	29
Engaged	1.00	0.94	0.97	36
Frustrated	1.00	1.00	1.00	34

The overall accuracy based on the confusion matrix is 96.75% for the four classes of engagement detection.

6. Conclusion

While comparing the models consisting of five and seven layers, the accuracy of the model consisting of only three layers is superior. As the number of layers in the model increases, there is a discernible trend toward a reduction in the precision of the representation. So, the model that has a lower total number of layers is the one that will provide the highest level of accuracy and is the one that is best suited for developing an image classification model. By looking at the graph of the training and validation accuracy of the three-layered model, we were able to determine that the accuracy of the five- and seven-layered models is not stable. On the other hand, in the three-layered model, it rises in an orderly fashion, and the final accuracy has the highest possible value. The analysis that was discussed before is reasonably near to explainable AI, which brings suitable comprehension for outputs provided by AI systems in computer vision. Explainability may be achieved in AI-involved systems, regardless of the domain, through the use of layered approaches, summarization, evaluation, and graphical comprehension, and this is consistent with the notion of ontology.

References

Ahmed, I., Jeon, G., & Piccialli, F. (2022). From artificial intelligence to explainable artificial intelligence in industry 4.0: A survey on what, how, and where. *IEEE Transactions on Industrial Informatics, 18*(8), 5031–5042. https://doi.org/10.1109/tii.2022.3146552

Alzubaidi, L., Zhang, J., Humaidi, A. J., Al-Dujaili, A., Duan, Y., Al-Shamma, O., Santamaría, J., Fadhel, M. A., Al-Amidie, M., & Farhan, L. (2021). Review of deep learning: Concepts, CNN architectures, challenges, applications, future directions. *Journal of Big Data, 8*(1). https://doi.org/10.1186/s40537-021-00444-8

Ashwin, T. S., & Guddeti, R. M. R. (2020). Automatic detection of students' affective states in classroom environment using hybrid convolutional neural networks. *Education and Information Technologies, 25*(2), 1387–1415. https://doi.org/10.1007/s10639-019-10004-6

Derave, T., Princes Sales, T., Gailly, F., & Poels, G. (2022). Sharing platform ontology development: Proof-of-Concept. *Sustainability, 14*(4). https://doi.org/10.3390/su14042076

Dhall, A., Joshi, J., Sikka, K., Goecke, R., & Sebe, N. (2015). The more the merrier: Analysing the affect of a group of people in images. In *2015 11th IEEE international conference and workshops on automatic face and gesture recognition, FG 2015*. Institute of Electrical and Electronics Engineers Inc. https://doi.org/10.1109/FG.2015.7163151

Dignum, V. (2021). The role and challenges of education for responsible ai. *London Review of Education, 19*(1), 1–11. https://doi.org/10.14324/LRE.19.1.01

Dwivedi, R., Dave, D., Naik, H., Singhal, S., Omer, R., Patel, P., Qian, B., Wen, Z., Shah, T., Morgan, G., & Ranjan, R. (2023). Explainable AI (XAI): Core ideas, techniques, and solutions. *ACM Computing Surveys, 55*(9), 1–33. https://doi.org/10.1145/3561048

Ehsan, U., Wintersberger, P., Liao, Q. V., Watkins, E. A., Manger, C., Daumé, H., Riener, A., & Riedl, M. O. (2022). Human-centered explainable AI (HCXAI): Beyond opening the black-box of AI. In *Conference on human factors in computing systems* -

proceedings. Association for Computing Machinery. https://doi.org/10.1145/3491101.3503727

Gade, K., Geyik, S. C., Kenthapadi, K., Mithal, V., & Taly, A. (2019). Explainable AI in industry. In *Proceedings of the ACM SIGKDD international conference on knowledge discovery and data mining* (pp. 3203–3204). Association for Computing Machinery. https://doi.org/10.1145/3292500.3332281

Hagras, H. (2018). Toward human-understandable, explainable AI. *Computer, 51*(9), 28–36. https://doi.org/10.1109/mc.2018.3620965

Holzinger, A., Saranti, A., Molnar, C., Biecek, P., & Samek, W. (2022). Explainable AI methods - a brief overview. *Springer Science and Business Media LLC, 13200*, 13–38. https://doi.org/10.1007/978-3-031-04083-2_2

Huang, T., Mei, Y., Zhang, H., Liu, S., & Yang, H. (2019). Fine-grained engagement recognition in online learning environment. In *ICEIEC 2019 - proceedings of 2019 IEEE 9th international conference on electronics information and emergency communication* (pp. 338–341). Institute of Electrical and Electronics Engineers Inc. https://doi.org/10.1109/ICEIEC.2019.8784559

Lavin, A., Gilligan-Lee, C. M., Visnjic, A., Ganju, S., Newman, D., Baydin, A. G., Ganguly, S., Lange, D., Sharma, A., Zheng, S., Xing, E. P., Gibson, A., Parr, J., Mattmann, C., & Gal, Y. (2021). *Technology readiness levels for machine learning systems.* arXiv. https://arxiv.org/abs/2101.03989v2.

Linardatos, P., Papastefanopoulos, V., & Kotsiantis, S. (2021). Explainable AI: A review of machine learning interpretability methods. *Entropy, 23*(1). https://doi.org/10.3390/e23010018

Lundberg, S. M., Erion, G., Chen, H., DeGrave, A., Prutkin, J. M., Nair, B., Katz, R., Himmelfarb, J., Bansal, N., & Lee, S. I. (2020). From local explanations to global understanding with explainable AI for trees. *Nature Machine Intelligence, 2*(1), 56–67. https://doi.org/10.1038/s42256-019-0138-9

Manasa, S., & Kumar, K. P. (2022). Digital forensics investigation for attacks on artificial intelligence. *ECS Transactions, 107*(1), 19639–19645. https://doi.org/10.1149/10701.19639ecst. Institute of Physics.

Mikalef, P., Conboy, K., Lundström, J. E., & Popovič, A. (2022). Thinking responsibly about responsible AI and "the dark side" of AI. *European Journal of Information Systems, 31*(3), 257–268. https://doi.org/10.1080/0960085X.2022.2026621

Purificato, E., Lorenzo, F., Fallucchi, F., & De Luca, E. W. (2023). The use of responsible artificial intelligence techniques in the context of loan approval processes. *International Journal of Human-Computer Interaction, 39*(7), 1543–1562. https://doi.org/10.1080/10447318.2022.2081284

Thiruthuvanathan, M. M., & Krishnan, B. (2022). Multimodal emotional analysis through hierarchical video summarization and face tracking. *Multimedia Tools and Applications, 81*(25), 35535–35554. https://doi.org/10.1007/s11042-021-11010-y

Thiruthuvanathan, M. M., Krishnan, B., & Rangaswamy, M. (2021). Engagement detection through facial emotional recognition using a shallow residual convolutional neural networks. *International Journal of Intelligent Engineering and Systems, 14*(2), 236–247. https://doi.org/10.22266/ijies2021.0430.21

Trocin, C., Mikalef, P., Papamitsiou, Z., & Conboy, K. (2021). Responsible AI for digital health: A synthesis and a research agenda. *Information Systems Frontiers*. https://doi.org/10.1007/s10796-021-10146-4

Wearn, O. R., Freeman, R., & Jacoby, D. M. P. (2019). Responsible AI for conservation. *Nature Machine Intelligence, 1*(2), 72–73. https://doi.org/10.1038/s42256-019-0022-7

CHAPTER THIRTEEN

Human AI: Social robot decision-making using emotional AI and neuroscience

Rumi Iqbal Doewes[1], Sapta Kunta Purnama[1], Islahuzzaman Nuryadin[1] and Nughthoh Arfawi Kurdhi[2]
[1]Faculty of Sport, Universitas Sebelas Maret, Surakarta, Central Java, Indonesia
[2]Faculty of Mathematics and Natural Science, Sebelas Maret University, Surakarta, Central Java, Indonesia

1. Introduction

Everyone has a general concept of what the term "consciousness" refers to, but it is best to avoid being too particular about characterizing it until we learn more. Until more is known about the nature of the problem, any attempt to provide a formal explanation is likely to be erroneous, limited, or both. This is because it is impossible to respond until you have a better understanding of the problem. Throughout history, most notable thinkers, religious figures, psychologists, and scientists have all devoted significant thinking and writing to the phenomenology of the human experience. According to recent cognitive science studies, this endeavor necessitates both physical and mental work (Alemi et al., 2014). Attitudes, beliefs, needs, and behaviors are just a few of the many manifestations of consciousness that come from this process. All of these could have their roots in the early stages of brain development. All these phenomena have their origins in the level of consciousness of everyone. We do not yet have a thorough understanding of this problem because, despite tremendous developments in domains such as computer science, neurophysiology, and brain imaging, scientific approaches have only been around for a short time (Bartneck and Forlizzi, 2004). For a long time, it was assumed that the concept of consciousness was invisible, immeasurable, and so outside the scope of science. This distinction is no longer held in high regard because it was discovered to be based on a fictional separation of the mind and the body, which was heavily affected by historical religious and cultural standards. The intellect and the body were clearly separated, and this separation was affected by

Emotional AI and Human-AI Interactions in Social Networking
ISBN: 978-0-443-19096-4
https://doi.org/10.1016/B978-0-443-19096-4.00013-4

cultural and religious concepts. Numerous studies have found a significant link between our physical bodies and our mental, emotional, and behavioral states. This theory is no longer being researched because a correlation has been shown. Trials on brain-damaged patients backed up his views, and the results enabled him to explain how human awareness evolves from feelings and emotions to a focus on the body. The fact that these novels are part of the same series as those listed earlier in the sentence strikes me as particularly relevant. In a previous argument, Dennett claimed that the formation of an internal network of mental representations was responsible for the emergence of intentionality. If a subject lacks some sort of autonomous system that allows him to distinguish between creatures in the same environment as he is, he cannot have intentionality, beliefs, or desires, and hence no consciousness. Without such a system, the individual is unable to distinguish himself from other objects in the same surroundings (Chen et al., 2020). The obviousness of this fact cannot be challenged, and there is no reason to do so. We propose using a strategy known as "knowing by generating" to accomplish this. Our goal is to keep these concepts relevant by applying them to the fast-growing field of social robots. The application of these features in the form of rules may determine a social robot's ability to learn and adapt sophisticated human-like behaviors. If a truly human-inspired control architecture is chosen, a new control paradigm for social robotics could be built. The architecture of this skyscraper would be sensitive to human nature. Some Potential Applications for Social Robotics and Cognition. Although there are numerous interpretations of the phrase "social robot," they all share important traits (Nair et al., 2021). According to these experts, social robots can take many different forms and perform a wide range of duties, but they must always be able to recognize people around them, engage in conversation with them, express their own artificial emotions, and comprehend those of their human counterparts. As a result, robots that can converse naturally with people are in higher demand. They must also be able to speak in a way that is like how people speak. As a result, you should be able to express yourself using body language, facial expressions, and other intuitive approaches in addition to words. This word is still applicable, but when a certain length of time has elapsed, it may no longer be. Over the last 10 years, the number of social robots on the market has surged, drastically expanding the spectrum of jobs that these machines may perform. As a result, the number of people who want to buy and use one of these devices has skyrocketed. It is clear that they are taking on more equal peer roles than the traditional servant role. Because of the nature of the change, their new role will gradually diverge from their previous one. To meet their demands, it is critical to work on attributes such as empathy, expression, and plausibility. When we analyze the cognitive system in charge of a particular robot's operation, we always find that one quality takes precedence over others. This is because the robot's brain ultimately governs its behavior. Because of the current technological state of these robots, we can witness huge breakthroughs in social robotics (Bosse et al., 2008). Kismet, the cartoon character featured in the following, is an excellent example of the type of cognitive system used in social robots. It is, in fact, one of the best-known systems of its kind. To enable Kismet, the social robot, to convey lifelike qualities, comprehend social behaviors, and communicate with humans, a modular architecture was developed. These objectives could only have been met with the assistance of this modular architecture. As a result, the robot was given the ability to duplicate these tasks on its own. Despite this, the major goal of the system was to simulate the interaction that develops between a baby and its primary caregiver. As a result, the robot was believable and expressive, but the agent was unable to adjust to different speakers or discern their emotions as the interaction progressed (Redstone, 2017). This resulted in a very convincing and realistic robot, but it also prohibited the agent from perceiving the feelings of the other person. Despite possessing a high level of realism, believability, and expressiveness, the robot was unable to perceive the other person's emotions during the conversation. This study used a new robot named Leonardo to improve these talents of "mind-reading" and "perspective-taking." In a 2006 study, the humanoid robot iCub gained high marks for its capacity to express emotion and communicate with others. It was made to seem like a youngster and has a body that is like that of an infant The ability to pursue and seize moving things is one of these talents. This is because the product's design considers regular human motion patterns (Matsusaka, 2012). Our research led us to the conclusion that there is a strong association between the complexity of a task and a certain method. Everyone in our organization has been fascinated by this subject for a long time. One approach is to divert attention away from sophisticated neural networks. However, many scientists today are developing cutting-edge technologies using a biomimetic approach. They concluded that this plan has a better chance of succeeding. These institutions are attempting to replicate human intelligence by studying and simulating the operation of various brain regions and neural networks (Mizera et al., 2019). This research was motivated by the desire to understand how human awareness develops. This is attributable to the fact that prior to engaging in self-monitoring, the brain selects content for worldwide broadcasting (C1; C2; C2). They say that even with all the gains in technology, most of the computations performed by modern machines are still, in essence, mimicking the unconscious processing (C0) that takes place in the human brain. And this is even though they have made some progress recently. The cognitive architectures we examined are all cutting-edge developments. However, none of these scenarios has shown how the agent's body and emotions generate personal preferences after learning about them. This type of advancement could 1 day serve as the foundation for an AI-powered brain. It is impossible to imagine a better solution to the existing dilemma. This cutting-edge cognitive architecture was designed to serve as an implementation of the Bosse computational paradigm. We will run the system through several tests to assess whether awareness has been obtained, as well as how consciousness has altered the artificial agent's beliefs and social behavior once it has been enhanced. After the system has been changed, we will proceed to this phase.

2. Background

The system separates the incoming data into those that are reliable and useful and those that are inconsistent and meaningless using a technique known as pattern matching. To reach this conclusion, a comparison of the incoming data and internal representations is performed. Implementing this procedure directly results in a significant reduction in the amount of time and energy required by the system (predefined templates). With the help of this information, the system can determine which parts of the data can be trusted to be accurate. This occurs after the fact is removed from the list of facts. To meet these standards, it will be necessary to either compile entirely new secondary data or modify the design of the existing templates. BRS is an abbreviation for behavioral rule sets (Cominelli et al., 2017). When these standards are implemented, they will alter how the templates react to and handle a wide range of situations. This set of rules is made up of the following sections, and the order in which they are presented reflects the relative weight assigned to each: Each of the three types of behavioral rule sets can be represented by one of the following acronyms: To extract a specific feeling from an emotional state template, the Feeling Rule Set algorithm is used. As a result of the progress made, we have reached the fifth step of the process. As a result of this investigation, a conclusion has been reached regarding a minor aspect of the agent's personality (Zaraki et al., 2017).

The SOmatic Marvel Rule Set(SOMARS) is currently the sixth most popular rule set in the entire world. A set of rules must be followed to characterize the somatic marker mechanism. When a labeled entity is recognized, the rules may examine the agent's emotional and physical state to determine whether the agent was experiencing any specific "feelings" at the time the entity was labeled. This is done to determine if the agent oversaw labeling the object. This is done to see if the agent was experiencing any particular "feelings" at the time the entity was labeled. This allows the rule to assert a somatic marker each time an entity to which it applies is identified (Dehaene et al., 2017). The rules may consider the person's current psychological and physiological state when deciding whether to require them to use a somatic marker. Alternatively, the rules may not consider the person's current emotional and physical well-being. The acronym "REAsoning Rule Set," or REARS, refers to a set of guidelines that allows the development of reasoning chains and the drawing of deductive inferences from the analysis of established facts to assert higher-level facts. To put it another way, REARS does not connect specific templates. The rules are referred to by the acronym "REARS." We will be able to represent expanded states of consciousness more accurately in our simulations with the help of these symbolic abstract reasoning principles (Cameron et al., 2017). As a result, we have highlighted this feature in the help window with a gleaming gold arrow. EXERS, or Execution Rule Set, is an acronym for a ruleset that is intended to be implemented at the very end of the procedure. As a result of our decision, we would not be paying them much attention or taking them seriously soon. Once all the other rule sets have finished updating the templates and a course of action has been decided, the EXERS will be able to issue instructions to the ACT Block so that those plans can be implemented. The step involves the robot's actuators receiving the motor commands generated by the robot control so that the robot animator can process them. These changes in the social environment are considered additional stimuli to the extent that the agent is the one who initially causes them (Thiessen, 2023). The body reacts to the subsequent physiologic changes as if they were a completely new stimulus that it had never encountered before. It is widely acknowledged and accepted that both external and internal stimuli originate in the environment that surrounds the organism. The structure found in the resources, which includes both examples and guidelines, is summarized in the following section. The emotion module includes many templates, including body preparation templates, reaction templates, external stimulus templates, and reaction templates. The Emotion Module also includes reaction templates. In addition to these, there is a group of templates known as reaction templates (Nair et al., 2021). The Emotion Module includes several subsystems and a small number of rules from REARS and EXERS in addition to the rule sets EMORS and STD-BEHRS. The following are the rule sets: You can find examples of the rules in the following categories in the Emotions Module, which you can access here: These are just a few of the features and elements that are included. Here are some additional characteristics and elements: The factors and elements discussed in this article may have an impact on an individual's feelings as well as their actions (Lazzeri et al., 2018). Given that multiple modules can use the same template file at the same time, this is not out of the question.

2.1 The importance of emotions in the formation of personal identity

We chose this strategy to follow in the footsteps of Bosse and his colleagues and replicate their success because their method was entirely dependent on installing the Emotion module into their SEAI system. As a result, we make it a priority to imitate their storytelling style as accurately as possible. The body preparing to take shape can be found in the software package itself, under the "Emotion" tab. The Emotional Circumplex Space (ECS) representation and its applications for interacting with other people's emotions are thoroughly explained in the SEAI framework documentation. The symbols for the two ECS point characteristics, valence (v) and arousal (a) have each been assigned a value of one, and the point has been assigned this value (a). The valence (v) of a point indicates whether it elicits a positive or negative emotional response in the observer, whereas the activation level (an) of a point describes the intensity of emotion felt (Mazzei et al., 2012). When attempting to describe an emotional state that is triggered by environmental cues and linked to a specific physiological state, the term "(v, a) point" is frequently used as a term of description. This specific emotional state frequently coexists with a recognized physiological state. The precise location of the source of this sensation in the body can be described as a "point." The coordinates contain a detailed description of the body in the state in which it is currently operating at peak efficiency (v, a). When an agent is confronted with an emotional stimulus, their instinct is to do whatever it takes to prolong the experience for as long as possible, scenario is true and the SENSE module of SEAI includes some essential tools for analyzing acoustic data (Nair & Bhagat, 2020). The agent in the given scenario has an emotional reaction caused by listening to music. As a result, we will be able to concentrate on the consequences of these assumptions. The representative within you will react to the music. For example, this program can analyze both the decibel level and the beats per minute (bpm) of the background music. As a result, we can conclude that SEAI occurs when: (i) the agent's microphones detect music as an external stimulus; The application's SENSE module processes audio after the user contributes some audio and generates a meta-map based on the data in the audio. This meta-map provides a detailed illustration of characteristics that are frequently associated with music. CLIPS for the I-Protocol and the Operating System After being transmitted to the brain, the meta-map is compared to a pattern derived from the representation of the music. This determines whether the music will be played (its tempo and volume). If the data do not match what was expected, the meta-map will be removed from the fact list; otherwise, it will be added (where one condition could be that the bpm is greater than 0). Music has been shown to activate the brain's Reasoning-Chain and Assertion Regions (REARS), which can lead to the presentation of truth claims and the ability to perform logical inferences. When a new musical fact is discovered, the EMORS will send out a warning signal to alert the other entities. Some musical compositions, such as EMORS, have the potential to cause physiological reactions when played at high enough volumes (Metta et al., 2010). Because the fact list containing the values (10, 1, and 2) is subject to an internal stimulus update, the value of the readiness shift changes from bp(0, 0) to bp as a direct result of this action. This has the immediate effect of changing the value of the previous condition, denoted by the letter bp (v,a). The coordinates are sent to the ACT block, and the EXERS activate (v, a). The ACT module will be available once the user arrives at the address indicated by the brackets in the sentence (v, a). Behavior is triggered when the appropriate musical cue and physical change occur at the appropriate time. The BEHRS rule that controls this behavior typically uses the bp (v, a) coordinate from the reactions template as its own. This type of behavior is common under BEHRS regulations. The two distinct accounts provide a wealth of information about both the music and the actual transition that occurred in the scene. The recipient can feel the emotion because the agent's outward display of emotion acts as a channel for the emotion to be transmitted to the recipient.

The third stage of the process is outlined in the third section of the process (LP0-LP1-LP2-LP3). Currently, the system is only capable of reacting in a specific way, processing data, and displaying the same emotions across all

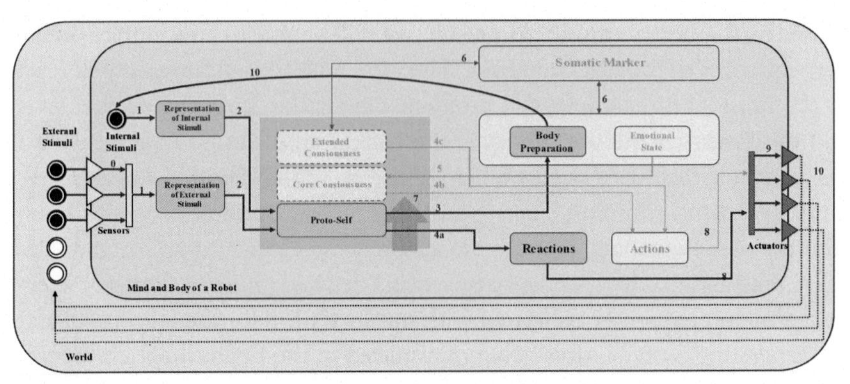

Figure 13.1 Social emotional artificial intelligence (SEAI) with only the emotion module loaded.

tests. It currently has only a few capabilities, which are quite limited (D'Arcy et al., 2011). When given the same input, the agent will always respond in the same manner, and its responses will never lag the input. When the agent is given the same stimulus, both claims are demonstrated to be true. Fig. 13.1 depicts the conclusion of the first, more fundamental stage of synthetic consciousness development.

3. The proposed method

Damasio defines emotion as a neural reaction to a specific stimulus that occurs involuntarily. Emotions such as happiness, sadness, apathy, or ambivalence can all be felt. When a person is in an emotional state, their body undergoes a series of changes that can be observed from the outside as emotional expressions. These physical changes occur when a person is in an emotional state. Victims of this sickness may have emotional symptoms. This is because connected brain activations usually comprise the behavior-planning process (Alharbi and Huang, 2020). Accepting the idea that emotion is the unconscious recognition of a specific physiological condition is the first step toward identifying an emotion. As a result of this realization, the organism has the sensation of feeling something. This realization occurs when an organism realizes that the internal model it employed to explain its own physical situation has altered. As a result of this understanding, the organism will now be able to perceive the sensation of feeling. To reiterate, as an organism grows, it becomes more aware of its surroundings and the activities that have taken place. Damasio divided the occurrence of these phenomena into five stages in his classification scheme. The following are some examples that fall into this category: an emotional stimulus that activates the organism, such as the visual system processing a specific item, causing mental representations of the thing in question to form second when the object is observed, the signals created by its visual processing activate brain areas that have been carefully designed to respond to cues relevant to the category to which it belongs (emotion-induction sites). Third, the brain's emotional centers oversee starting a chain reaction in other brain regions as well as the body, releasing the entire range of body and brain reactions that are responsible for emotional experience (Westbrook and Braver, 2015). Fourth, both the subcortical and cortical parts of the brain use first-order neural maps to account for changes in the body's condition. There will come a point when you feel exceedingly sensitive. The pattern of brain activity that occurs in parts of the brain critical for emotion formation is mapped using fifth- and second-order neural structures. Other neuronal structures, including those in the second and third levels of the hierarchy, are involved in the mapping out of alterations to the proto self. The word "emotion object" refers to the behaviors that people conduct in response to emotional triggers in their lives. Even the systems' fundamental dynamics might be successfully replicated. Damasio's opinions are likely to have inspired the strategy's conceptual roots. These dynamics can be viewed in several ways, one of which is as a series of progressively complex states. There are many other points of view to consider besides this one. The term "states" refers to the great variety of neurobiological states caused by the different activities that occur in the brain. During the data abstraction process, the following techniques and tactics were used (Kashyap, 2021). A higher level of precision can be obtained by modeling the many states and activation patterns of the brain as discrete states. In some cases, the state property in question is recognized by a single letter. This is achievable if the values of the states are distributed in a specified way. All this preliminary work is done to have a better understanding of the modeling technique. People work together to accomplish this. This is done to ensure that everything is entirely understandable and open. When seeking to characterize the dynamics of the systems they were studying, Bosse and his team typically used the passage of time as a critical reference point. A person's dynamic traits derive from their ability to reflect change through time, which also includes their ability to connect states that occur at various times. This was done to gain a better grasp of the underlying dynamics at work in these processes. We projected that the agent's actions would be influenced by their mental state in this situation. The technique is described using the word "LP," which stands for "local dynamic properties" and is derived from the LEADS TO notation. The proposed approach can be used if desired. For the active sensory representation of hearing music, the internal state property (music) is investigated here, along with a vector of prepared state properties p = (p1, p2, ...) for body reaction activation. This is done to facilitate the discourse. In this case, both considerations are considered. This is a composite state attribute with several dimensions, indicated by the notation S = (p1, p2). Fig. 13.2 clearly depicts the technique, and the LPs that go with it are as follows. When assessing the status of a sensor, the performance of music in LP0 time can be used as a useful comparison. The following formula can be used to compute the LP1 sensor state (music) SR: S is more potent than p LP2 SR (music).

Damasio's writings were first made available to the public during these years, particularly in 1994 and 2000. The evolution of Antonio Damasio's theory of mind over time has been likened to the construction of a massive skyscraper. Damasio developed his theory of mind to shed light on the mechanisms underlying consciousness formation. According to him, an individual's emotions come first, then their feelings, and finally their "feelings of feelings." They made the proto self, the center of consciousness, and the expanding awareness out of the materials they had on hand begins with introspection, progresses through increased self-consciousness, maintains expanding one's sense of center, and culminates in increased awareness of

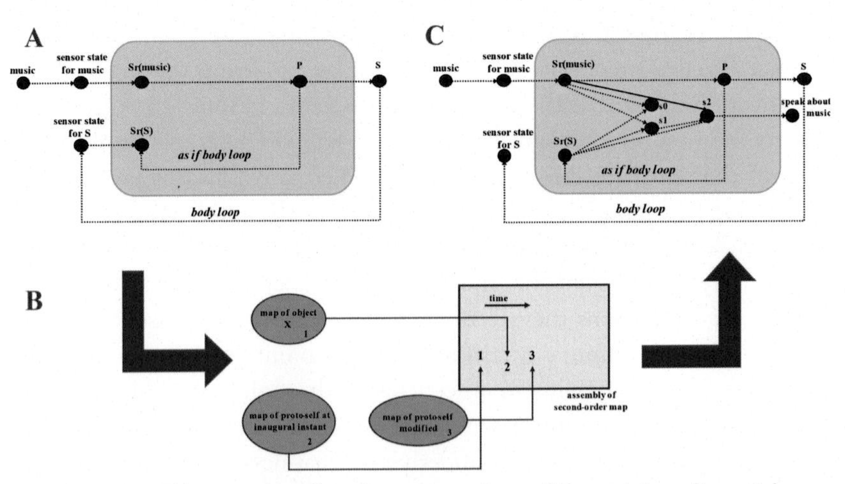

Figure 13.2 This essay heavily references sections of Damasio's earlier articles.

one's surroundings. These three layers are components of the human body's complex network of connections. This second part, however, should be viewed as a tool that is an essential part of the process of making new discoveries rather than as the location where this technique is put into practice (Saerbeck et al., 2010). Temporal trace language, abbreviated as TTL, is one temporal language that can be used to convey this type of information. The entries in this dictionary cover historical eras, time periods, and historical trajectories that are beyond a reasonable doubt. When someone refers to their list of visited states as "a voyage through a state," they are referring to the order in which they have visited them. They created a scenario in which an agent listens to music, changes their emotional state, and then changes their behavior because of the experience using this descriptive model. Our overarching assumption was that a participant's mental state would influence their behavior. The letter "LP" in LEADSTO stands for "local dynamic properties," a concept developed at the time the notation was created. This one sentence could sum up the entire strategy. If you wish, you may carry out the recommended course of action. In this chapter, we look into the possibility of eliciting a physiological response by combining the properties of a "prepared" state (p = (p1, p2, ...)) and an "internal" state. This research will improve our understanding of the sensory representations used by musicians in their work. When listeners participate in this way, they can better process what is being presented to them. In this situation, these two factors were considered and discussed. Fig. 13.2 depicts the approach in greater detail, and the following LPs are commonly used in conjunction with it: Surprisingly, there are many striking parallels between how sensors work and how music is performed in LP0 time. You can use the following formula to determine what the LP1 sensor is doing when music is playing: The relative activity of p LP2 SR is noticeably lower when compared to S, which has a noticeably higher value.

These offerings are made up of a lot of different apps that cooperate with one another while also carrying out a variety of separate tasks. A piece of open-source middleware called YARP was created with the goal of simplifying the creation of decentralized platforms for controlling robots. This base is necessary for everything, including the delivery of services and the transfer of data across networks. The green components, which are service-specific, are called modules. Data collection (from sensors or the network), data processing (using the collected data), and data transmission are all handled by modules (to the next node in the network). When discussing the transmission of data from one component of a system to another, the term "XML packets," which refers to a serialized form of structured data objects, should be used because it is the most accurate and appropriate of the available options. SEAI can expand your business as you add new products or services because of its scalability and adaptability to a variety of deployment models because it is very flexible. With the aid of practically any computer system or programming language, such a service can be created. Now that SEAI has flexible data management, this is a real possibility. The flexibility of its services allows SEAI to meet the needs of a wide range of customers. The proposed architecture includes some parts with the labels "ACT," "SENSE," and "PLAN," but these are all merely filler. The reactive subsystem can be used as a metaphor to help understand the symbolic connection between the SENSE and ACT services.

3.1 Power generation methods that can be applied to electrical applications

Within this robot, sensor is the actual energy meter. This service oversees controlling the robot's voltage and current levels as well as how they are connected to the power source it uses. The robot's power consumption in watts at 1 hertz can be precisely measured by the power supply monitor (PSM) service. Finally, the data collected by the PSM are used to create reports on the health of the robot after being serialized so that it can be transmitted over a network (proprioception). The robot's actuation system, of which this service is a part, is crucial to the operation of the robot. Because the system translates the instructions from the language in which they were originally written, the animators can understand the deliberative system's instructions. Modules that were created specifically for use inside the robot's body have been installed (e.g., hands, arms, neck, and face). The HEFES is a great example of such a component (hybrid engine for facial expressions synthesis) (Scassellati et al., 2012). Our earlier research described a technique for programming the facial robot's emotions, and this technique was used. The robot animator is given an ECS point at the coordinates, and this module generates a set of servo motor settings for it to use to simulate the expression (v,a). Regarding the model, an ECS has been assigned to this component. More details on meta-scene objects like this one can be found in the SENSE section of this guide. This dataset contains a list of individuals along with their assigned unique identifiers and locations (x,y,z). When the time is right, the robot will receive a signal from the deliberative system identifying the person or object that the system wants it to focus on. Additionally, the YARP port is constantly monitored by the Gaze control module for data of this kind. When it discovers such information, it will react and act appropriately.

3.2 The robot animator was employed to produce animated robots

The first action required before the robot can move is to enable the low-level service. Robot control is constantly asking for this service, which also includes head and face movement. Due to its innate concurrent behavior, the robot can complete requests that conflict with one another. This is because the algorithms establish the causal connection between the two. Even though the training is delivered on time, this still occurs. The PLAN block, which controls the brain's in-circuit longitudinal predictive staging, enables communication between the reactive and the deliberative layers of the system. This section is frequently shortened to "Interactive CLIPS." The I-CLIPS Rules Engine was developed because of using CLIPS to improve earlier work the full description of this engine can be found. The fact acts as the primary organizing principle for rules in expert systems like CLIPS. The bits and bytes that make up information are represented visually in these charts. Each individual item being recorded after a count must have a new entry made in the ledger's tally column. A group of labeled fields is referred to as a "template" by I-CLIPS, and these fields are used to store defined data. Declarative languages use templates, whereas procedural languages use objects. Both procedural and declarative programming languages frequently employ templates and objects. By using this specific technique, objects can be transformed into I-CLIPS examples. The concept of a template in declarative languages serves a function like that of an object in procedural languages. The process of making decisions begins with a review of the applicable criteria. All the LHS requirements must be met for the rule to go into effect. Calling functions, adding data, and modifying templates are just a few examples of RHS operations. Data are generated when new facts are asserted and can be used to inform rulemaking or shared with other networked services. When already-existing facts are revised, these data can also result. The term "feeding," which is a common noun that describes the process, is frequently used to describe the act of imparting one person's knowledge to another. It is true that a rule is activated when its left-hand condition is met, but this happens before the rule is put into action. A list of all the rules that are currently in force is placed on the table and is arranged so that it can be read from top to bottom in the order that it should be disabled. Always proceed according to the agenda's suggested order. Although I-CLIPS modules are used in this instance, CLIPS modules are more frequently used to refer to the modules in question. The sets of documents that are specific to each module contain the rules and file formats for that module, a file that includes a graphic representation of the module in addition to the module's code (.clp). The associated rules and templates can be defined and saved within the SEAI Knowledge Base after a module has been successfully imported into the I-CLIPS Rules Engine. What sets a module apart from other components of its kind is the function that it performs. Each node in the network can simultaneously receive, process, and transmit data and is free to do so on its own schedule. Multiple modules can accept the same incoming data at the same time, as shown in Fig. 13.3. The information presented by the Meta-Scene is currently being absorbed by both the Attention Module and the Emotion Module. Altering the Energy Module's internal parameters can be done for two different reasons: either to broadcast new data to the network, as is the case here, or specifically to broadcast no data at all. The parts of the human brain that handle processing emotions

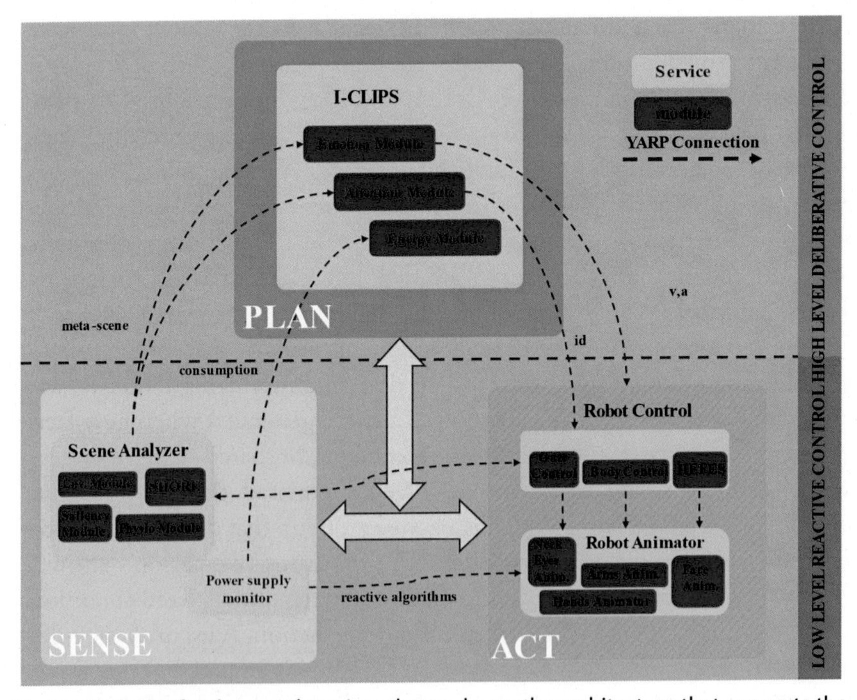

Figure 13.3 The fundamental services that make up the architecture that supports the social emotional artificial intelligence (SEAI).

and paying attention are shown in Fig. 13.3. The modular design of SEAI is one of its distinguishing features. One of this design's distinguishing features is the user's ability to add or remove entire modules and then activate or deactivate features even as the program is running. This could be viewed as the main objective of SEAI's work. Before a module can make use of functionality provided by another module, it must first create a dependency on that module. Rules of Conduct for Module B You might be able to include checks on the status of the templates that are defined in module A, for instance. Let us say, for the sake of clarification, that the B module included both the A and C modules. The inaccessibility of Modulus B if Modulus A has not been loaded is irrelevant to the topic at hand because it is not a necessary condition. Both actions are included in the loading of a module process, so the terms are equivalent. Given this information, you should proceed with a healthy dose of caution.

Setting priorities and using "dummy facts" are two techniques that, when used together, can produce excellent results when it comes to time management. The methodology described here is compatible with other methodologies found in the body of prior research because it adopts a causal approach. Unfortunately, we only have a very little amount of control over the passage of time. The significance of the rules should be established right away because doing so can help their resolution move up the priority list. Employing the word "priority" is one tactic for achieving this goal. It is possible to quantitatively evaluate the importance of the rules in the context of discussions about the real world and the laws that govern it. $10,000 is one potential cost estimate. The activated rules that are also the most relevant will take precedence on the list of actions to be taken. Saliency will default to 0 if no value specification is provided. You can stack the rules that are related to a module using this technique. A layer, which is also known as a sub-module, is created when two or more templates are linked together in accordance with the same ruleset. For convenience, every rule is contained in the same system sub-module. In the business world, a set of rules is referred to as a "Rule Set." T2 must be changed if T1 is altered; however, this cannot be done the other way around. The rules from the same set will appear on the timeline in the order determined by the conflict resolution method if there is a conflict. The rules from the same set will appear in the order they were decided upon if there is no conflict. The fact that CLIPS offers users a variety of options for achieving calmness in their work can be partially credited to the system's adaptability. Due to the striking similarities between it and human thought processes, the depth strategy ultimately prevailed. By definition, a depth strategy gives precedence to the initial activation and final deactivation of a fact-based rule. Even after the rule is no longer in force, this is still true. As a direct result, people start acting in ways that are more consistent with the current situation and are influenced by it. Use the term "dummy facts" as an alternative if you are looking for one. This latter approach relies on assertions of fact to guarantee that the rule sets are executed in the proper order. These procedures will enable us to ensure that the rule sets are applied in the proper sequence. The system was given the nickname "dummy" because, whenever a question was asked of it, the data that were relevant to it was immediately removed from the list.

3.3 Porting the computational model in the SEAI framework

A model of the process that produces our somatic markers will also be discussed. Because it was left out of Bosse's original design, this model is being presented to the public for the first time. To begin, a unique graphical representation of the data management procedures used by the SEAI Cognitive System must be created. The AI of the SEAI stems from its structure, which is like the human brain. The organizational patterns that allow the various functional components to work together efficiently comprise the system's structure. When we talk about something's "structure," we are really talking about the connections that make up that thing. The SEAI's database will be gradually expanded with modules describing the three levels of awareness; these modules will serve as benchmarks and guidelines for the AI's actions. This framework treats our system's sensory states as facts or templates, whereas rule sets are dynamic properties at the local level. Fig. 13.4 shows

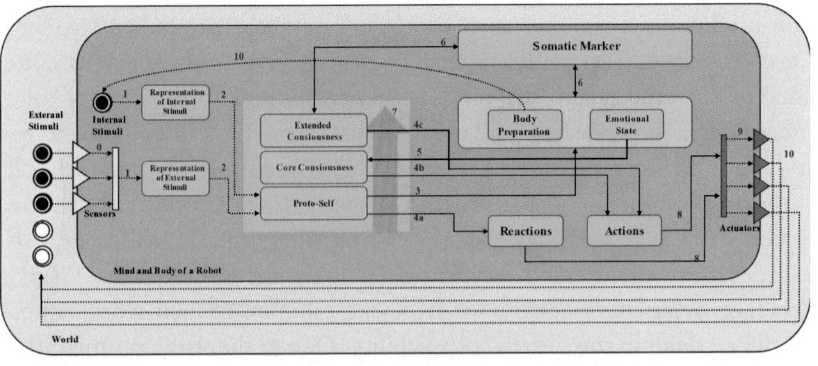

Figure 13.4 Our attempt to Port Bosse into social emotional artificial intelligence (SEAI).

the SEAI cognitive system in its entirety, from the loaded modules to the rest of the hardware.

When the numbers highlighted in Fig. 13.4 are examined more closely, the information presented in the following paragraphs becomes apparent. The SENSE node collects, filters, and combines data acquired through one's senses after it has been transmitted through it. It processes data from various sensors. To make sense of everything that has happened, the deliberative system analyzes the meta-maps after receiving data via exteroception or interception (through the process of interoception). We appreciate you taking the time to read this. We will be able to provide a more understandable and easier-to-understand explanation if we use specific examples. The fifth chapter delves deeper into the question of where humanity stands in relation to the rest of the universe. Fig. 13.4 depicts the robot submerged in water while enclosed in a gray box that represents the real world. The white box and the gray box with the dividing line in the middle are both meant to represent the robot's chassis. A robot's "senses" gather information about its surroundings, while its "actuators" move objects around. The robot's eyes are a group of triangles arranged in the middle of the empty space that makes up the robot's body. These triangles cover the area where the robot lives. In this diagram, the yellow and red triangles represent the network's incoming sensors and outgoing actuators, respectively. In other words, the figure depicts how the robot's perception system may detect some external stimuli but not others. The fact that the robot can detect some external stimuli demonstrates this (represented by the dark blue circles). Stimuli that may be relevant in the context of social robotics include object properties, human traits, and environmental factors. You could classify these various types of stimuli as "stimuli." This category also includes the characteristics of the objects under discussion (like temperature, noise level, or luminosity, for example). Environmental factors include temperature, background noise volume, and lighting intensity. In most cases, a perception module designed specifically for that purpose will receive the raw data collected by each sensor, process it, and analyze it. This is like the initial sorting that our senses do when we take in information about the outside world. To continue along these lines, the motor system requires the actuation system of an artificial agent to function properly. There are many different types of actuators available, with servomotors accounting for most installations. Each group of motors housed within a robot's body is controlled by a different animator; each animator operates independently of the others. They oversee the entire situation. Actuators, also known as "drive units," serve the same purpose. Actuators are mechanical devices that control movement. Because there are so many options on the market, you should have no trouble finding an actuator that meets your needs. These could be the lights that give the appearance that your skin is blushing, or they could be the speakers used in speech synthesis. Actuators create directional arrows, which are then used as a proxy for the robot's actual subsequent movements. The agent will reacquire the modifications to their environment caused by the robot's actions and store them in their memory as a new set of cues.

3.4 Uninterruptible solar power system

The internal structure of the PLAN block is depicted in Fig. 13.3 after it has been stretched to its maximum length. Before this model was created, the PLAN block was stretched to its maximum length. Now, this specific aspect of the model is receiving most of the attention and emphasis that is being directed toward it. The ACT and SENSE nodes combine perception and action services into a single offering. The services related to perception are depicted in yellow, while the services related to actuation are depicted in red. This article contains numerous instances where this diagram and Fig. 13.3 are strikingly similar. The arrows represent the various rules that can be applied in the situation, and the blue boxes can be viewed as examples. Using the layered approach discussed thus far, arrows representing causal and temporal relationships can be drawn. One method is to draw these arrows in layers. The mental and physical state of an agent at any given time can also influence how the agent reacts to stimuli from the outside world. A blue circle represents the agent's internal state variables in the diagram. In either case, these values may serve as a representation of the agent's mental and physical well-being. If they made the necessary changes, either of these two options would suffice. At the end of each execution cycle, the agent's internal stimuli are generated anew. This iteration considers the most recent information processed from the agent's internal and external environments. The single gray square in the middle, which is made up of three different layers, can be used to summarize the entire image. This allows you to see the image in its most natural state. The "fact list" is stored in the robot's "working memory," also known as the "gray area" of its memory. The agent keeps a "fact list," in which it records all its knowledge about the world and itself. There are details about both the world and the individual. This includes information from both inside and outside the organization. The three symbolic levels can be interpreted as a metaphor for how one's mind develops because of being exposed to increasingly abstract knowledge (Ramirez-Asis et al., 2022). Learning new things leads to a natural expansion of one's mind. This progression is the result of being exposed to increasingly complex informational formats. This mental state can be developed simply by consuming increasing amounts of abstract information. This level of consciousness is attainable, and as you encounter more ethereal truths over time, it will develop and become more firmly established. This level of consciousness is attainable and will eventually become more firmly established. A YARP connection to several services or other types of connections is not always indicated by arrows pointing in the same direction. To put it another way, this is not a hard-and-fast rule. After reading these explanations, you should have a better understanding of the nuances associated with rule sets and modules.

3.5 Primary self-awareness and emotions

Because of the addition of the Feeling Module, new rule sets and templates with an azure hue can now be created. This is illustrated in the fifth figure. Throughout this lesson, you will see the emotional state template. According to recent research into the inner workings of the mind, mental rehearsal may be distinct from physical practice. In contrast to bp, which produces an instant response, es(v,a) is a system internal parameter that the system uses to modify the robot's behavior. Unlike blood pressure, which has an immediate response, this happens gradually (v, a). bp, on the other hand, is an external parameter (v,a). This is true (v, a), even though BP (v, a) points still designate qualitatively distinct states. Following its investigation, the FEERS will conclude that the individual's current emotional state is a fact that should be added to the list. It must be in this configuration to function properly. If you follow these instructions in the correct order, the actions template will perform an action and insert the values (v' and a') associated with $v'=(k1)$ wherever they are needed. The global variable k, which stands for the influence factor, is accessible to all modules. Its effect on the agent is determined by its value, which can be any value between 0 and 1. $*vbp + k*vesa'=(k1)*abp + k*aes,v'=(k1)*vbp + k*vesa'=(k1)*abp + k*aes$ the following is an example of what K is capable of:

Until then, nothing will change. There are three different modifications that can be made to bring the ES values into compliance with the most recent version of the EMORS regulations. According to established conventions, ves increases when the music's tempo quickens and aes decreases when the volume is too low. Furthermore, due to the following rule, v increases as the musical tempo increases: Both conditions are observed as rules concurrently. Consider the common practice of listening to soothing music at a constant volume for long periods of time. Long-term exposure to this genre of music has been linked to the side effects, so listeners should proceed with caution. The FEERS will declare the fact es(v,a) true; the REARS and ALT-BEHRS will analyze the fact using the actions template and enter the values (v,a); and the FEERS will declare this. As the run progresses, the FEERS will report that es (v, a) is true, with ves and aes decreasing in value. Because the agent currently has a BP and an ES in her working memory, the ALT-BEHRS will be triggered. So, as soon as we can, we will put it into action. The Feeling module's EXERS rule definitions are given more weight than the module's contributions to the EXERS rule set. As a result, the Feeling Module generates the EXERS rules, which are then verified by the Actions Template. To put it another way, if it is known that the values for both actions and reactions have been established, and if the values of the actions and reactions are known ahead of time, the outcome can be predicted. This is because the subsequent condition is the same as the previous condition because the underlying declarative process is unaware of the services provided by the ACT block. As the experiment progresses, the Feeling Module will cause the subject's calm expression to change to one of boredom. Despite his or her initial lack of response, the subject is likely to become bored as the experiment progresses. This should be expected as a natural outcome of the experiment. When the biological model's subcortical level is raised to the cortical level or when LP4 and LP5 are added to the computational model, the results are very similar. When subcortical activity equals cortical activity, these two procedures can be used. These two procedures can be performed once the subcortical tissue has been raised to the level of the cortical tissue. In other words, this is a metaphor for how one feels when a feedback loop in their own body begins to work. This way of thinking will eventually result in more sophisticated mental images than the original one. These symptoms are more difficult to detect early on because they are caused by gradual and subtle changes in the body. Both our actual emotional experiences and our mental representations of them change all the time. It would be premature to bring up music because agent has yet to show any discernible reaction to any evocative object. Damasio's "core consciousness" is depicted in Fig. 13.5, and it appears to be functioning normally. The way the participants' emotional receptivity and reactions to the music changed throughout the experiment demonstrates this.

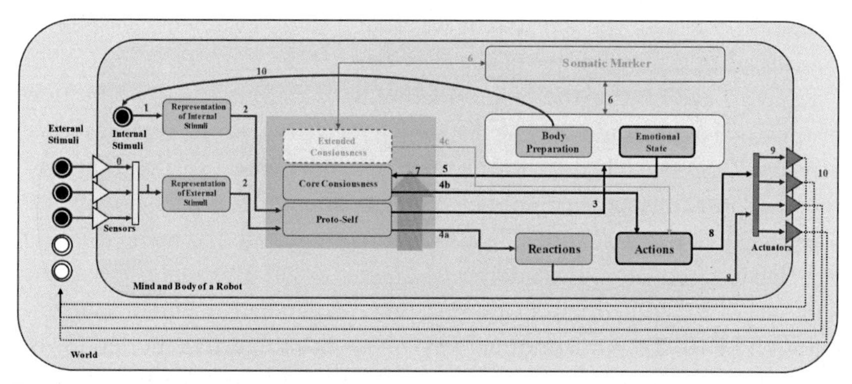

Figure 13.5 Social emotional artificial intelligence (SEAI) after feeling module loading. New parts highlighted in blue.

The subjects' increased self-awareness demonstrates this. Expanding one's awareness, which is another phrase for "one's own emotions," is comparable to one's own emotional experience. Our study and subsequent research were inspired by Damasio's "somatic marker hypothesis," which contends that improving one's physical condition can improve one's mental health and receptivity (Mohanakurup et al., 2022). Somatic markers are the links between a trigger and an emotional response. These connections are formed by the body in response to changes in a person's emotional state. In fact, SMs use their physical bodies to form sentimental convictions and assessments about the world around them. This process, which is aided by SMs, is largely responsible for our consciousness' development. Our perceptions and judgments of those things can influence how we interact with them. When we have enough information, we can approach these entities using our assumptions, suppositions, and opinions. If we are later exposed to the marked entity, whether the marker has been removed, this mechanism will cause us to remember our physical state at the time it was activated. This will influence how we perceive and react to the marked entity in the present. Bechara et al. used data from "Iowa Gambling Task" studies on both healthy and damaged brains. The primary goal of these studies was to determine whether the SM mechanism helps reduce gambling-related financial losses. Participants' brains were either healthy at the time of the studies or had been injured before taking part. The SOMARS were created as a physical representation of the mind-body connection. To learn more about this aspect of our brains, we conducted a preliminary computational experiment. The experiment used a computer-generated version of the Iowa gambling task as its foundation. Because the old system was found to be insufficient, it will be gradually phased out in favor of the new one.

Keep an eye out for the blue underlining, which indicates where new information has been added. The diagram depicts the Somatic Observational Memory Rules, which can also be abbreviated as green arrows. The first thing that probably comes to mind is a remark about the music being played at the time (for example, "this music is getting boring"). If a marked object or person is discovered, this additional outcome is also possible. Using the digits "4c," "8," and "d," in that order, one can generate a sequence beginning with LP10. To put it another way, an agent's subjective models (SMs) are dynamic and self-created; they change because of interactions with objects in their environment. This biologically inspired mechanism is activated by the FOF module, resulting in the agent's biassed behavior, warped worldview, and formation of autobiographical memory. This demonstrates that the FOF module oversees activating this mechanism. Outside forces, for example, could have had the opposite effect on the agent's emotional state and put him in a better "mood," and the outcome could have been different. This could have had a different outcome. This is because the agent's chances of success would have been higher if they had begun in a more favorable "mood." This is the reason for what occurred. The term "chill-out music" refers to a type of music that aims to lift listeners' spirits and encourage a state of peaceful introspection. Chill-out music is also known as "relaxing music." This is correct because it conveys the impression that someone is relaxed. This stage of consciousness is not possible without the stages of consciousness that came before it because it transcends the "here and now," includes subjective assessments of concrete objects in the world and allows for the conception of more abstract ideas. It should come as no surprise that previous stages of consciousness are required for the one currently being experienced because this level of consciousness has been reached and implies that all previous levels of awareness have been resolved. Fig. 13.6 depicts the extended consciousness stage in its activated state. In the following experiment, a humanoid robot named FACE with an SEAI brain implant was used. Because of its lifelike facial mask, which can mimic a wide range of human expressions, the robot has an uncannily similar appearance to that of a human. To give the impression that the android is more natural than it is, 32 servomotors are used to move the Frubber material on the mask, which translates to "flesh rubber." The mechanical system of the robot also includes a head that can move in different directions and eyes that can look in different directions. Because the mannequin's head has

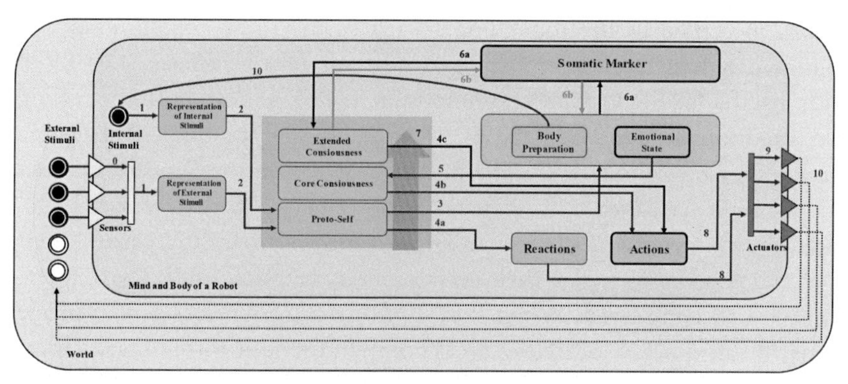

Figure 13.6 The proposed approach.

been attached to its body, which is seated in a chair, it is now acting as a passive research subject. A simulated workplace is one of the tools used during the human resource interview. There are no props or changes to the environment to make the HRI appear more authentic; it is simply a typical office.

The investigation can be divided into four distinct subfields, which are listed below: The show begins with the persona ID1 entering the space occupied by the robot. He does not greet the robot, ignores his immediate intrusion into its personal space, does not respond when he does, stands there with his arms folded for a few moments, and then walks away. His actions, such as those mentioned above, all fall under the category of being rude or inconsiderate. In the second scene, ID2 enters the space and begins acting in a variety of ways, including greeting the robot, invading its personal space while also taking a step back, chatting with the robot for a while, and then leaving the space. The final action in this series is ID2 leaving the room. When ID2 leaves the room, the chain of events ends. In the third scene, ID3 enters the room, politely greets the robot by grinning, strikes up a lengthy conversation, and then exits. The scene is now finished. When it is finally time for him to leave, he gives her one last greeting before saying goodbye. For the next minute and a half, they remain motionless, avoiding any movements that might alert the robot to their presence. Then, without exception, they all leave the location of the event. Kinect Studio, a piece of software that can record and playback depth streams, color streams, and audio from a Kinect, was used to record this sequence as a reproducible scenario (Parashar et al., 2022). This was done to ensure that it could be used again in the future. As a result, the same social situation can be presented to the robot under each of the three conditions of the cognitive system listed below to evaluate their relative effects: The SEAI with just the feeling and attention modules, the SEAI with both the feeling and attention modules and the SEAI with just the attention and feeling modules are conditions 1–3. The Scene Analyzer on the Kinect examines the images it has taken and estimates the number of important social cues based on the people in the scene. The distance between the parties, their ages, and genders, as well as their gestures, body language, and facial expressions, are all considered. When the SENSE service receives a new image frame, it can quickly locate the focal point and send it to the user. This is accomplished using unaltered image analysis based on a variety of factors such as colors, contours, light contrast, quick motion, and many others. This location also has its own special ID, which is represented by the number. When there is nothing particularly interesting happening in the scene, the robot will analyze it by focusing on the feature that stands out the most. This happens when there is nothing visually appealing in the frame.

4. Results

Memory is currently being loaded with the "Emotion" Module. As a result, the facial expression changes, causing an emotional response to the activities taking place in the body's social environment. When the FACE algorithm determines that there are no humans in the field of view it is monitoring, it displays a sad face with a low level of arousal on the screen (between −0.3 and −0.5). Fig. 13.7 depicts two concurrent outcomes. To be more specific, we came to the following conclusions: The Emotion module's rules will cause a change in the robot's physical state, and the Attention Module's rules will cause the robot's attention to be redirected away from the salient point and toward the identified subject. This will be done in response to the identified salient point. This physical change will be communicated using the robot's FACE expressive capabilities, which translate a point on the ECS into 32 commands for the robot's face and neck servomotors. This change will be announced on the robot's face. This is based on our theory of emotions, which holds that changes in one's physical condition are the source of one's emotions.

In this sentence, each of the four points of view is referred to as a "column." Each row contains, in that order, the winner's ID, bp (v), and bp (a). The second is the standard unit of measurement for time, which is commonly denoted by the letter "t." Fig. 13.7's bp(v) chart depicts the

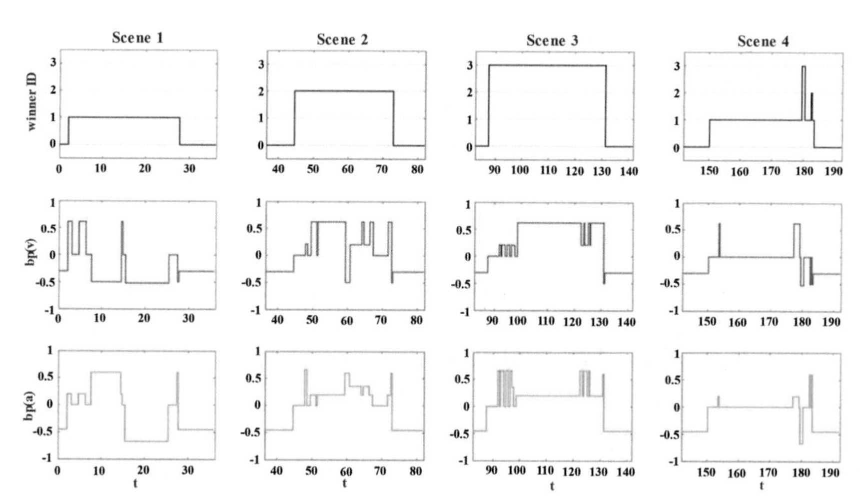

Figure 13.7 HRI's findings from the combined FACE and social emotional artificial intelligence (SEAI) experiment (condition 1).

robot's emotional response, while Fig. 13.7's bp(t) chart depicts its mental state. Fig. 13.7 depicts this comparison (a). In this case, simply listening in as an observer is perfectly acceptable. When someone invades FACE's personal space, she becomes uneasy (0.5 and 0.6) and angry (0.52 and 0.67). She smiles (0.21 and 0.6) when someone greets her or smiles at her, and she shows interest when they speak to her (0.62 and 0.2). Even though the subjects' specific actions while interacting with the robot are not revealed, the pattern of BP (v, a) demonstrates how the robot is emotionally affected in the first three scenes. Case ID1's rude behavior is largely responsible for the first person's abnormally high number of negatively valenced values and wildly fluctuating levels of arousal. During a fascinating conversation, the robot introduces himself to ID2 as impartial and courteous. The result is an increase in positive valence and a decrease in anxiety. The robot's interaction with ID3 generated a plethora of uplifting stimuli, resulting in an endless stream of happiness and excitement in the robot. This was due to the numerous motivating cues provided throughout the interaction. Scene three shows how the interaction impacted the robot. The pivotal scene demonstrates how remarkably resilient the robots' detection abilities are in the face of population variation. The three test subjects do nothing but stand motionless and silent in front of the robot. However, once the subjects leave the room, the robot's mood returns to neutral (0, 0). People are identified as they exit the building through multiple exits at the same time, causing changes in the number of people leaving to coincide with this change. The overlap will make it difficult for the scene analyzer to piece together the facts. The information has been filtered, so even if a sudden or significant change occurs, the robot animator will not allow the robot to respond to it. The attentive model assumes that during the first three scenes, FACE will concentrate intently on just one thing. This subject should either fill the entire frame or be the primary focal point. And each of the three scenes I just mentioned fits the bill (ID 0). Even in the absence of social cues, a robot's attention is focused on what is truly important. This occurs both before and after subjects are made aware of it. The final scene includes all subjects, but the robot is only interested in subject ID1. Subject ID1 is ideal for this experiment due to his proximity to the robot and the fact that no other subjects are actively attempting to attract the robot's attention. Even though environmental factors are clearly to blame for FACE's declining physical health, the agent is still unaware of its own mental state. Each emotion lasts exactly if the initial stimulus. We have no idea what happened if anything happened at all. As a result, FACE does not publicly endorse any positions and meets only when necessary (Nair & Bhagat, 2020). Whether or not the test subjects return to the area, the robot will continue to focus on the people in the area. Here is the evidence I have gathered to back up my claim. In many ways, the proto-historical state of consciousness and the current state of consciousness is similar.

Along with the feeling component, a mental state template is automatically loaded. Fig. 13.8 depicts how the features of this module affect water ripples. The FEERS module can perform emotional enchantment after the FEERS module has effectively sown the knowledge seeds in the agent's working memory (s). Invading the robot's personal space has a significant effect on him between the 10- and 15-second marks in scene one but has no effect on him between the 100- and 130-second marks in scene 3. The robot's expression changes as a result. When an agent's internal representation of emotions changes, behavior patterns, both variable and absolute, can change (VBP and ABP, respectively). Section 5 delves deeper into the weighted influence factor, denoted by the notation ($k = 0.1$), and the revised values, denoted by the notation (v, a). The old values had to go to make way for the new ones. Because ALT-BEHRS is the first command run, the commands are run in the order (v', a') rather than (v) (vbp, abp). If one's emotional state becomes more stable and larger over time, the gap will close. As a result, the distance between them shrinks. The es and

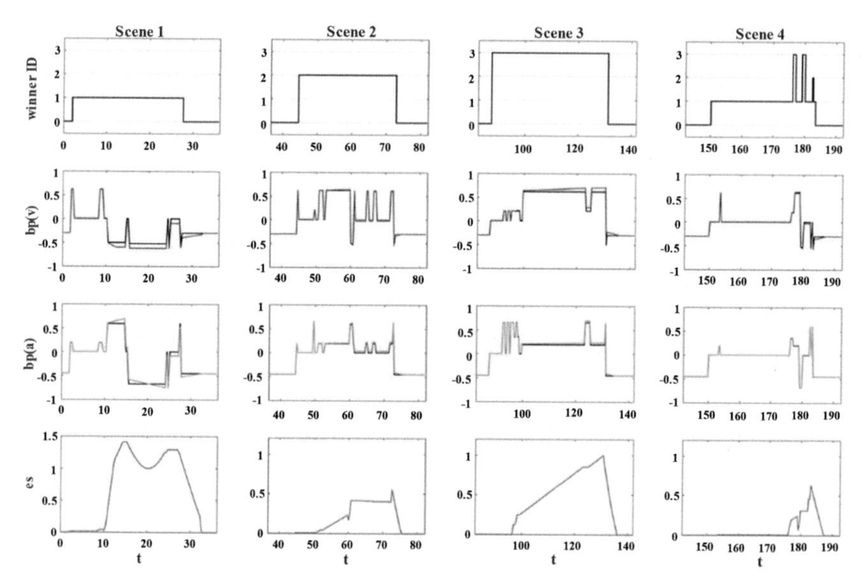

Figure 13.8 The following are the results of HRI's experiment in condition two using FACE integrated with social emotional artificial intelligence (SEAI): Each of these four distinct points of view is referred to as a "column."

bp in scene 3 after t = 130 show that the transition from subject detection to loneliness is not always as smooth as it appears, and the es trend can last much longer than the causative stimulus. The BP pattern after t = 130 is yet another indication that the emphasis has shifted from subject detection to loneliness. One example of this idea in action is the relationship between subject detection and increased feelings of isolation.

These emotions are entirely fabricated by me. As a result, the agent is aware of the simulated feelings it is currently experiencing. Emotions also appear as a continuum of the agent's physical manifestations of emotional states, which can take many different forms. This is what people mean when they say, "physical expression of emotions." In any case, all this knowledge is transient, and while it does temporarily alter behavior, it is only transient. The agent's intentional behavior does not change significantly because of the agent's inability to recall the emotions that drive specific actions. They continue to dominate the FACE popularity rankings, where ID1 is currently ranked first. Attempting it for the third time, the incorporation of the FOF module allows SEAI to use the somatic marker mechanism, resulting in the creation of the SOMARS definition. This is the situation as a direct result of the SOMARS definition. Fig. 13.9 depicts

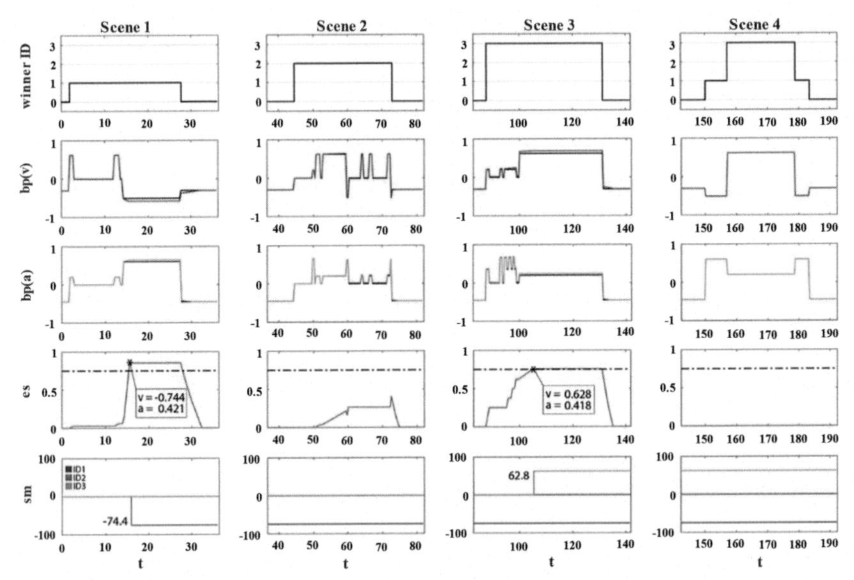

Figure 13.9 The HRI-conducted FACE-integrated social emotional artificial intelligence (SEAI) study yielded information about condition three test.

the experiment's results in relation to this third configuration. When Scene four is performed in front of each subject, there is a noticeable change in the robot's attentive behavior, more consistent emotional reactions, and the absence of any discernible emotional state. This should come as no surprise given that it is a direct result of the mechanism for the creation and recall of SMs discussed in subsection 5.6. For the purposes of this experiment, we have determined that $s = 0.75$ is the desired level of sensitivity. Because of ID1's annoying behavior, the es intensity rapidly increases and surpasses the s threshold at $t = 15.5$ s. This is a direct consequence of the behavior. When the game is rerun (at $t = 15.83$ s), an SM will be generated with the winner ID, the current score, and a marker value of 74.4 calculated using the formula in subsection 5.6. This SM will be generated when the game is restarted. If you restart the game, this SM will be generated automatically. In the third scene, FACE is seen interacting with ID3, but this time the marker quality is better. This exchange occurs after the third scene. Somatic markers are released when an entity disturbs an emotional state. This is due to the agent's complete understanding of the marked entity—both what it is and how it should make them feel. This is since each subject is in front of the robot at the same time. This is because all of the test subjects are currently positioned in front of the robot. As soon as labeled subjects ID1 and ID3 enter the agent's field of view, the robot recalls the previous causative interaction's somatic state, which was either 0.5 or 0.6. This occurs as soon as the subjects enter the agent's field of view. When it comes to directing attention, our behavioral model shows that the rules associated with SPEC-BEHRS for positively marked entities take precedence over those associated with SPEC-BEHRS for negatively marked entities. This is because entities marked positively receive more attention than entities marked negatively. As a result, the robot is concentrating all its efforts on locating ID3 until it is confident that it has also located ID1 and ID2. FACE is particularly interested in him because of his reputation as an exemplary citizen and the organization's favorable feelings toward him. The face also has a positive attitude toward him. Because his interactions with the robot have never been harmful, the robot's emotions have never been strained to the point of breaking. As a result, during the pivotal scene, ID2 can vanish from the robot's consciousness. The robot did not gain enough new knowledge from this experience to necessitate the development of a specialized SM.

Each of these four distinct points of view is referred to as a "column." The following information can be found at the end of each row: Winner ID, Epsilon, Sigma, and Beta. This ensured precision. This was done to ensure the highest level of precision possible. You can learn the meaning of each color by placing them in the appropriate locations on the SM chart. The second is the letter "t," which represents the accepted unit of measurement for time. Damasio's SEAI system simulator was put through its paces in this final test, complete with a preset somatic marker mechanism. The agent has progressed to the point where it can now generate facts about the people and places in its social environment, which it can then memorize for later use. This indicates that the agent has achieved a significant goal. These are memories of how you felt, and the bodily sensations associated with those emotions can still be felt today. The following interactions have the potential to overtly or covertly bias behavior and have an impact on the agent's somatic state. Finally, a few final thoughts and observations.

5. Conclusion

The goals of this article are to present and discuss a novel cognitive architecture that can be used by social robots. We concluded that modeling a popular theory about the human mind as the cognitive system that controls the robot and then implementing it was the best way to create an expressive and emotional robot. We came to this conclusion after realizing that if you want to build an expressive and emotional robot, modeling should come first. After much deliberation and consideration, it was decided that the theory of mind would be modeled and used. This would be done to make the architecture compatible with other social robots. Because the architecture is modular, accomplishing this goal is not impossible. The modular layout of the architecture allows for this adaptability. Because they are not required for the system's intelligence, any of the agent's qualities may be transferred to another agent or changed in a way that is distinct from the others at any time. This can be accomplished without jeopardizing the agent's "personality," which includes their memories, beliefs, experiences, and behavioral traits. Furthermore, the artificial agent's ability to use inference reasoning is not constrained in any way. As a result, the cognitive block can be used to solve any problem that the artificial agent may encounter. The number of rules as well as the degree of complexity and specificity attached to those rules determine these constraints. Experiments demonstrated how SEAI programmed artificial feelings and emotions into a social humanoid that was influenced by its environment. Because the agent used these feelings to form judgments about the social environment in which it found itself, it was perceived to have a higher level of social competence. In the most recent experiment, for example, the robot deviated noticeably from its usual pattern of behavior. This experiment makes no claim to provide proof that we have created a sentient being. Contrary to popular belief, we did not create a sentient being. Contrary to popular belief, the following list includes only a few of the events scheduled to take place soon: Extending SEAI to include capabilities that have been identified as lacking in competing robotic cognitive systems. To gather answers to questions two and three, it will be necessary to conduct a survey in which participants rate how much they believe the robot is conscious. From this vantage point, the contributions of experts in behavioral psychology and neuroscience would be especially valuable. The most important consideration is whether the social robot's interactions with people will improve because of the changes to its regular routine. According to our theory, incorporating SEAI into this new class of robots will make them appear more realistic. As a result, the likelihood that people will accept and believe in them will increase. To sum up, we believe SEAI has the potential to be a successful approach for simulating human consciousness and, ultimately, a promising first step in addressing the question of whether robots can have synthetic forms of consciousness. We believe SEAI has the potential to be a useful technique for simulating human consciousness. As a result, we believe SEAI has the potential to be a viable method for simulating human consciousness. To summarize, this is our position. In the second scenario, ethical considerations will be of the utmost significance and importance.

References

Alemi, M., Meghdari, A., Ghanbarzadeh, A., Moghadam, L. J., & Ghanbarzadeh, A. (2014). Effect of utilizing a humanoid robot as a therapy-assistant in reducing anger, anxiety, and depression. In *2014 second RSI/ISM international conference on robotics and mechatronics (ICRoM)*. https://doi.org/10.1109/icrom.2014.6990993

Alharbi, M., & Huang, S. (2020). An augmentative system with facial and emotion recognition for improving social skills of children with autism spectrum disorders. In *2020 IEEE international systems conference (SysCon)*. https://doi.org/10.1109/syscon47679.2020.9275659

Bartneck, C., & Forlizzi, J. (2004). A design-centred framework for social human-robot interaction. In *RO-MAN 2004. 13th IEEE international workshop on robot and human interactive communication*. https://doi.org/10.1109/roman.2004.1374827. IEEE Catalog No.04TH8759.

Bosse, T., Jonker, C. M., & Treur, J. (2008). Formalisation of damasio's theory of emotion, feeling and core consciousness. *Consciousness and Cognition, 17*(1), 94–113. https://doi.org/10.1016/j.concog.2007.06.006

Cameron, D., Fernando, S., Collins, E. C., Millings, A., Szollosy, M., Moore, R., Sharkey, A., & Prescott, T. (2017). You made him be alive: Children's perceptions of animacy in a humanoid robot. *Biomimetic and Biohybrid Systems*, 73–85. https://doi.org/10.1007/978-3-319-63537-8_7

Chen, H., Park, H. W., Zhang, X., & Breazeal, C. (2020). Impact of interaction context on the student affect-learning relationship in child-robot interaction. In *Proceedings of the 2020 ACM/IEEE international conference on human-robot interaction*. https://doi.org/10.1145/3319502.3374822

Cominelli, L., Carbonaro, N., Mazzei, D., Garofalo, R., Tognetti, A., & De Rossi, D. (2017). A multimodal perception framework for users emotional state assessment in social robotics. *Future Internet, 9*(3), 42. https://doi.org/10.3390/fi9030042

Dehaene, S., Lau, H., & Kouider, S. (2017). What is consciousness, and could machines have it? *Science, 358*(6362), 486–492. https://doi.org/10.1126/science.aan8871

D'Arcy, L. P., Sasai, Y., & Stearns, S. C. (2011). Do assistive devices, training, and workload affect injury incidence? Prevention efforts by nursing homes and back injuries among nursing assistants. *Journal of Advanced Nursing, 68*(4), 836–845. https://doi.org/10.1111/j.1365-2648.2011.05785.x

Kashyap, R. (2021). Breast cancer histopathological image classification using stochastic dilated residual ghost model. *International Journal of Information Retrieval Research, 12*(1), 1–24. https://doi.org/10.4018/ijirr.289655

Lazzeri, N., Mazzei, D., Cominelli, L., Cisternino, A., & De Rossi, D. (2018). Designing the mind of a social robot. *Applied Sciences, 8*(2), 302. https://doi.org/10.3390/app8020302

Matsusaka, Y. (2012). Speech communication with humanoids: How people react and how we can build the system. In *The future of humanoid robots - research and applications*. https://doi.org/10.5772/27135

Mazzei, D., Lazzeri, N., Hanson, D., & De Rossi, D. (2012). Hefes: An hybrid engine for facial expressions synthesis to control human-like androids and avatars. In *2012 4th IEEE RAS and EMBS international conference on biomedical robotics and biomechatronics (BioRob)*. https://doi.org/10.1109/biorob.2012.6290687

Metta, G., Natale, L., Nori, F., Sandini, G., Vernon, D., Fadiga, L., von Hofsten, C., Rosander, K., Lopes, M., Santos-Victor, J., Bernardino, A., & Montesano, L. (2010). The ICUB humanoid robot: An open-systems platform for research in cognitive development. *Neural Networks, 23*(8–9), 1125–1134. https://doi.org/10.1016/j.neunet.2010.08.010

Mizera, C., Laribi, M. A., Degez, D., Gazeau, J. P., Vulliez, P., & Zeghloul, S. (2019). Architecture choice of a robotic hand for deep-sea exploration based on the expert gestures movements analysis. *Robotics and Mechatronics*, 1–19. https://doi.org/10.1007/978-3-030-17677-8_1

Mohanakurup, V., Parambil Gangadharan, S. M., Goel, P., Verma, D., Alshehri, S., Kashyap, R., & Malakhil, B. (2022). Breast cancer detection on histopathological images using a composite dilated backbone network. *Computational Intelligence and Neuroscience, 2022*, 1–10. https://doi.org/10.1155/2022/8517706

Nair, R., & Bhagat, A. (2020). Healthcare information exchange through blockchain-based approaches. In *Transforming businesses with bitcoin mining and blockchain applications* (pp. 234–246). https://doi.org/10.4018/978-1-7998-0186-3.ch014

Nair, R., Vishwakarma, S., Soni, M., Patel, T., & Joshi, S. (2021). Detection of covid-19 cases through X-ray images using hybrid deep neural network. *World Journal of Engineering, 19*(1), 33–39. https://doi.org/10.1108/wje-10-2020-0529

Parashar, V., Kashyap, R., Rizwan, A., Karras, D. A., Altamirano, G. C., Dixit, E., & Ahmadi, F. (2022). Aggregation-based dynamic channel bonding to maximise the performance of wireless local area networks (WLAN). *Wireless Communications and Mobile Computing, 2022*, 1–11. https://doi.org/10.1155/2022/4464447

Ramirez-Asis, E., Bolivar, R. P., Gonzales, L. A., Chaudhury, S., Kashyap, R., Alsanie, W. F., & Viju, G. K. (2022). A lightweight hybrid dilated ghost model-based approach for the prognosis of breast cancer. *Computational Intelligence and Neuroscience, 2022*, 1–10. https://doi.org/10.1155/2022/9325452

Redstone, J. (2017). Making sense of empathy with sociable robots: A new look at the "imaginative perception of emotion.". *Social Robots*, 19–38. https://doi.org/10.4324/9781315563084-2

Saerbeck, M., Schut, T., Bartneck, C., & Janse, M. D. (2010). Expressive robots in education. In *Proceedings of the SIGCHI conference on human factors in computing systems*. https://doi.org/10.1145/1753326.1753567

Scassellati, B., Admoni, H., & Matarić, M. (2012). Robots for use in autism research. *Annual Review of Biomedical Engineering, 14*(1), 275–294. https://doi.org/10.1146/annurev-bioeng-071811-150036

Thiessen, R. (2023). Social Robots to encourage play for children with physical disabilities. In *Companion of the 2023 ACM/IEEE international conference on human-robot interaction*. https://doi.org/10.1145/3568294.3579985

Westbrook, A., & Braver, T. S. (2015). Cognitive effort: A neuroeconomic approach. *Cognitive, Affective, and Behavioral Neuroscience, 15*(2), 395–415. https://doi.org/10.3758/s13415-015-0334-y

Zaraki, A., Pieroni, M., De Rossi, D., Mazzei, D., Garofalo, R., Cominelli, L., & Dehkordi, M. B. (2017). Design and evaluation of a unique social perception system for human–robot interaction. *IEEE Transactions on Cognitive and Developmental Systems, 9*(4), 341–355. https://doi.org/10.1109/tcds.2016.2598423

CHAPTER FOURTEEN

Human AI: Neurodegenerative disorders and conceptualization of cognitive ability

G. Maheswari and H. Indu
Department of Education, Avinashilingam Institute for Home Science and Higher Education for Women, Coimbatore, Tamil Nadu, India

1. Introduction

Neurodegenerative disease is an umbrella term for several disorders that primarily affect neurons in the human brain. It is a disease in which the cells of the central nervous system stop working or die. Neurodegenerative diseases pose a serious threat to human health.

Sigmund Schlomo Freud (1856–939) is the founder of psychoanalytic theory that explains human personality in depth. It is a method of analyzing unconscious conflicts based on personal associations, dreams, and fantasies. It is also a method of treating mental illness in psychology. Freud is also called the father of modern psychology. We are now in the 20th century known as Freud's time. In fact, he is an Australian neurologist who bridges the realms of science and literature as well as literary criticism. He believed that scenes from his childhood had a profound impact on his adult life. It shapes our personality, behavior, and cognitive development. A person's past traumatic experiences can be hidden from consciousness and emerge in adulthood. These age-related disorders have become more common in recent years, in part due to the growing geriatric population. Examples of neurodegenerative diseases include.

- Alzheimer's sickness (advert) and other dementias
- Parkinson's ailment (PD) and Parkinson's sickness
- Prion disease
- Synucleinopathies

Emotional AI and Human-AI Interactions in Social Networking
ISBN: 978-0-443-19096-4
https://doi.org/10.1016/B978-0-443-19096-4.00003-1

- Motor neuron disorder (MND)
- Huntington's sickness (HD)
- Spinal-cerebellar movements ataxia (SCA)
- Spinal muscular atrophy (SMA)

2. Literature review

Many scholars who have written papers and papers on creative writing and daydreaming. An article by Vijay Chavan of Maharashtra on Freud: creative writing and daydreaming. A Research Paper on Freud and Creative Writing: An Analysis of Writing, Dreams, and Dream Interpretation by Nirjarini Tripathi at Teresina College, Mysore. Sophie I's Essay on the Correlation between Dreams and Educational Environments. Lindquist and John P. McLean from the University of Queensland, Australia. Blog Article on Creative Writing and Dreaming - Freud - Dr. Critique and Theory of S. Srikumar.An article by Joseph Sandler and Anna-Marie Sandler on unconscious fantasy, identification, and projection in creative writing.

2.1 Creative writers and dreams

In Creative Writers and Dreams, Freud seeks to uncover the secrets of being a creative writer, where to tackle material, and how to navigate it in such a powerful way. They may doubt their own resources but fail at times and are unable to provide satisfactory or clear explanations. He identified these socially constructed selfish and erotic desires as the driving force behind fantasies and daydreams. Everyone is a true artist when they indulge in spontaneous ideas. Creativity is equated with conscious work, with the capacity for imagination to demand the ego.

2.2 The unconscious

Freud viewed aesthetic touch as the ability to retain artistic pleasure in the deeper pleasures of unconscious matter. Through an unconscious process, the artist was able to come into contact with private fantasies and the effects of public art. Fantasy is defined as "a creation of the imagination, whether expressed or merely imagined." Freud believes that fantasies are created by unfulfilled desires already stored deep in the unconscious. When a person starts living in a fantasy world, he will not worry so much about reality. These fantasies are created by the imaginary world as a dream to get out of their closed world and return to the public world. Freud himself, in his

"New Introductory Lecture on Psychoanalysis," explained the argument of his essay as follows: The dream is as follows. According to Freud, both dreams and forms of mental illness are products of repressed and unconscious impulses. This dream testifies to your mental anguish. This requires the analyst to interpret a person's dreams and reveal to him the causes of his obsessive-compulsive symptoms and their "transient nature" which is the main evidence of the truth and value of psychoanalysis.

2.3 Child's play

From Child's Play to Fantasy Freud explores a common factor in humans: the desire to change the existing unsatisfying world of reality. A child's favorite activity is games. This is not to say that all children act like creative writers. The artist creates his world or rather transforms it in a new way. However, the child does not take the world seriously, takes the game very seriously, and throws a lot of emotions into it. This space is occupied by mental activity that is aimed to creating a situation, where our unsatisfied wishes are turns into fulfilled.

2.4 Overview of other writers

Freud saw the presentation of human spirit in the form of art. He established the principles that evaluate the art within the preview of psychoanalysis. He followed the practice of literary sources like fantasies and daydreaming, and also applied his theory psychoanalytic tools into the study of literature. Alfred Kazin says, "He (Freud) brought, as it were, the authority of science to the inner prompting of art, and thus helped writers and artist to feel their interest in myths, in symbols, in dreams was on the side of reality, of science, itself, when it shows the fabulousness of the natural world." Plato calls poets mad, but Freud explains that artists are not mad but are dissatisfied with their limitations. The unconscious serves as a repository for desires and thoughts stored here and mediated by the preconscious. This stream of consciousness is a comprehensive conceptualization of cognitive processes, including dreams, memories, images, and more.

2.5 The Artist's unfulfilled desires

Freud says that an artist's choice of content seems to be driven not only by recent triggers but also by unfulfilled childhood desires. His past and present extend into the future through his writings. The artist's dream is spoken in front of the public. He realizes his own dreams by formally controlling the sources of our pleasure. Freud deals with the poetic influence of the content of a work of art. He calls this aesthetic response a "bribe" that overcomes disgust and frees one from anxiety. People's fantasies are harder to observe and understand than children's play. The child plays alone and makes physical engagement with others for the purpose of play. However, adults are ashamed of their fantasies and hide them from others. These fantasies are explained by personal motives. A happy person never daydreams, only dissatisfaction. The driving force of fantasy has to do with a fully determined and unsatisfying reality. Depending on the gender, the motivation for this desire may be different. It can provide an outlet for the artist's creative imagination.

2.6 The process of dreaming

In the process of dreaming, the artist engages in formative work that is far from the archetype of dreaming. The artist sits in his mind and looks at other characters from the side. Thus, through self-observation, he is able to divide his ego and personify the current conflicts of his mental life. The dreamer carefully protects his fantasies from others because he feels that he has reasons to be ashamed. But when a creative writer presents his work, we take great pleasure in his personal dream. This is called "The Poetics of Lies" in the art of overcoming barriers and emotions between each one.

2.7 The satisfaction of repressed desires

Freud explores the nature of literary imagination and aesthetic enjoyment through a unique synthesis of the creative writer and the dreamer. Essentially, Freud sees the creative writer's function of leading dreams as "the fulfillment of repressed desires." He treats the author as an egoist, dressing infant fantasies into useable adult forms. Most dreams are indirect and symbolic acts of repressed desire. During the awakening process, the superego does not allow these personal needs to enter consciousness. During sleep, the unconscious mind camouflages itself and manifests itself as a dream. Repressed desires are indirectly realized in dreams. Creative artists maintain a certain space for independence to express their material choices and variations. However, the dreamer puts aside his fantasies and never wants to express them because he is ashamed of his desires. In his work Creative Writers and Dreams, Freud identifies reality for works of art. Don't be intimidated by the complexity of formula. I actually think it will turn out to be too sparse a pattern. However, this view of creativity may not be fruitless. I will never forget that he emphasizes childhood memories in the artist's life. We must not seek to return to the various kinds of creativity, which must be perceived as reworking of existing materials rather than being original.

2.8 The relationship between the writer and the dreamer

Freud focuses on comparing the novels of the writer and the dreamer, not the classics. He says that one of the common characteristics of all these works is the central character or hero. The hero's journey is the ego of both the author and the reader. He suggests that creative writing tasks are the content of children's play. He presents a comparative version of the creative writer and the dreamer in his work "The Interpretation of Dreams" "Sleep is the liberation of the spirit from the oppression of external nature, the liberation of the soul from the shackles of matter". "Dreams reveal the deepest and most secret desires that the average person cannot express. Real feelings often have no place in everyday life and, according to Freud, are expressed through dreams. A person may have too many desires due to the pressures of life, and dreams may arise to suppress his thoughts and feelings. In this work, Freud focuses on learning about the unconscious by watching all these dreams. Creative writers have a flair for literary genres and can put their dreams to good use. "Many people waste their dreams because they don't know how to live them." However, creative writers can sublimate their dreams into literature and awaken our sensibilities. He does not consider the fantasy to be the end of unfulfilled desire. Finally, Freud tries to answer his own question about what creative writers do and how they evoke emotions in us. It is said that dreamers suppress their fantasies because they are ashamed of them. Even if they did, other people would be asking them questions or answering them. Freud exclaims why we get a damned pleasure from a creative writer's presentation. He says we can only speculate on how this happens but cannot draw conclusions.

3. Pathology

In medicine, the etiology of a disease or condition refers to regular examinations aimed at determining more than one factor that together cause a disease. Therefore, epidemiological studies examine whether relevant factors such as location, gender, exposure to chemicals, and many others increase or decrease the likelihood of a disease, condition, or illness in a population once a disease has become widespread. Certain genetic changes that increase the likelihood of the disease have been identified in some cases. Environmental factors also contribute to the development of neurodegenerative diseases. For example, there is evidence that Parkinson's disease is associated with long-term exposure to pesticides, toxins, and chemicals. About 5 cases in 2016 were reported. Four million Americans have Alzheimer's disease. It is estimated that by 2020, 930,000 people in the United States will be living with Parkinson's disease. Several brain diseases have a significant impact on cognitive impairment in the elderly.

3.1 Clinical and anatomical classification

Clinical symptoms are determined by the affected system and do not reflect the molecular pathological background. In most cases, the symptoms overlap or converge during the course of the disease. Therefore, the clinical classification is mainly used for the evaluation of early clinical symptoms. The main clinical symptoms of neurodegenerative diseases are.

3.1.1 Decreased cognitive function, dementia, and changes in higher brain function

The main anatomical regions involved were the hippocampus, entorhinal cortex, limbic system (amygdala, olfactory cortex, anterior cingulate, subcortical structures), and neocortical areas. Focal cortical symptoms may include focal degeneration of the frontal, temporal, parietal, or occipital lobes. A subtype of dementia is frontotemporal dementia, which includes frontal and temporal lobe degeneration (FTLD). These patients have behavioral or speech problems. It is important to distinguish between rapid and slow forms of cognitive decline.

3.1.2 Movement disorders

Clinically, they are associated with symptoms of immobility or ataxia/rigidity, such as the so-called parkinsonism. Hyperkinetic movement disorders such as chorea, dystonia, ballistic disease, atherosclerosis, tremors, tics, and myoclonus; ataxia; or symptoms of upper and lower motor neuron damage. In many cases, a combination of these symptoms can be observed in some forms of the disease, both at the onset and during the clinical course.

4. Anatomic distribution of neurodegeneration

"Frontotemporal lobar degeneration is a group of distinct disorders characterized by neurodegenerative changes affecting the brain. Frontotemporal lobar degeneration is caused by progressive damage and loss of nerve cells in the frontal and temporal lobes of the brain." In most people, this is accompanied by a build-up of one or other of two proteins, tau or TDP-43. In FTD, these proteins are misfolded (malformed), which causes them to accumulate inappropriately in brain cells, interfering with or destroying the normal functioning of these cells. Clinical subtypes of FTD can also be classified as "tauopathy" or TDP43opathy, depending on how the misfolded protein accumulates in the brain. A third protein, FUS, accumulated in place of tau or TDP43 in approximately 10% of cases. Accumulation of tau or TDP-43 can also be seen in other neurological disorders. The clinical manifestations of frontotemporal lobar degeneration are varied. Affected individuals may experience gradual changes in behavior and personality and may have difficulty thinking and communicating effectively. The progression and the specific symptoms that occur can vary from person to person. In general, the clinical symptoms of these disorders can be broadly grouped into three categories, manifesting changes in behavior, language, and/or motor function. "Extrapyramidal disorders: refer to all disorders or dysfunctions of the extrapyramidal system, which groups together several areas of the brain interconnected by complex neuronal circuits." They combine three main symptoms: resting tremor, akinesia, and hypertonia. Since Parkinson's disease is the most characteristic form of the extrapyramidal syndrome, extrapyramidal syndromes are wrongly associated with it and the term "Parkinsonian syndromes" is often used to refer to extrapyramidal syndromes in general. However, there are other disorders with about the same symptoms that do not correspond to Parkinson's disease. They are called non-Parkinsonian extrapyramidal disorders. Non-Parkinsonian extrapyramidal disorders usually result from dystonia, a motor neurological disorder characterized by a disturbance of muscle tone and manifested by involuntary and prolonged muscle contractions that cause abnormal attitudes.

- Acute dystonic reactions
- Akathisia
- Pseudoparkinsonism
- Tardive dyskinesia

5. Factors contributing to neurodegeneration

Hundreds of neurodegenerative diseases exist and, with few exceptions, their origins are unknown. Moreover, even when the etiology is identified, the exact cellular mechanisms that initiate disease and cause neurodegeneration often remain speculative. Several factors can cause neuronal degeneration, many of which are linked to neurodegenerative diseases.

5.1 Genetic causes and risk factors

There are several rare neurodegenerative diseases that have a clear genetic cause. Mutations disrupt the function of genes essential to neuronal or glial cell biology and lead to the early onset of serious neurological diseases. Other genetic variants carry risk factors, meaning they increase the likelihood of idiopathic neurodegenerative disease without directly causing it.

5.2 Epigenetics

Altered epigenetic regulation, such as altered DNA methylation, modification of histones, or epigenetic regulatory pathways and enzymes, although rarely directly responsible, may be implicated in the development of many neurodegenerative diseases.

5.3 Toxic substances

Toxic substances, including alcohol or lead, can cause neurodegeneration, especially with prolonged exposure. Toxins can directly cause neuronal cell death or impair neuronal or glial function, resulting in neurodegeneration.

6. Protein misfolding and impaired protein clearance

A subclass of neurodegenerative diseases, called proteinopathies, is associated with protein misfolding and the formation of aggregates, often in the form of Lewy bodies, neurofibrillary tangles, or plaques characteristic of the corresponding disease. The accumulation of proteins and the formation of aggregates leading to proteotoxicity are considered to be one of the fundamental mechanisms underlying the pathology of these neurodegenerative diseases. Protein aggregation can be caused by protein variants that are more prone to aggregate formation, ubiquitin-proteasome defects, autophagic clearance of lysosomal proteins, or, in the case of prion diseases, a protein mal folded transmissible Forms that can cause normal variants of a protein to misfold the same protein.

6.1 Altered cellular signaling

Aberrant cell signaling, such as disrupted presynaptic inputs or weakened intracellular signaling pathways, may contribute to the pathogenesis of neurodegenerative diseases, especially since neuronal viability and survival often depend on the synaptic activity and extrinsic survival signal.

6.2 Altered energy metabolism

Neurons are highly metabolically active cells: neuronal signaling directs high levels of protein synthesis and high energy demands, and many transmembrane transporters and pumps require ATP input. Thus, neurons are very vulnerable to minor disturbances in energy metabolism. They mainly depend on an optimal supply of oxygen, glucose, and lactate from blood, astrocytes, and glial cells. Mitochondrial dysfunction has fatal consequences, leading to oxidative stress or apoptotic cell death.

6.3 Oxidative stress

The high density of mitochondria and high levels of activity in most neurons lead to the formation of large amounts of reactive oxygen species (ROS). The regulation of oxidative stress by antioxidants is essential for neurons. Disruption of superoxide dismutase or glutathione peroxidase is associated with many neurodegenerative diseases. Additionally, oxidative stress can be induced by inflammation or impaired glial function and lead to Neurodegeneration.

6.4 DNA damage

Accumulation of DNA damage and decreased efficiency of DNA repair mechanisms are associated with general dementia and cognitive decline as well as various neurodegenerative diseases.

6.5 Altered cytoskeleton and axonal transport

Neurons are physically large cells with most of the cytoplasm located quite far from the perinuclear cytoplasm and all of its metabolic machinery. Cellular function and health are highly dependent on a mechanism that distributes cells along axons, called axonal transport. Metabolites such as proteins and lipids, vesicles, and organelles are transported bidirectionally along the axon. This transport can be passive but most often relies on active transporter proteins and coupling proteins that pull cargo along the axonal cytoskeleton. Disruption of transport mechanisms or the cytoskeletal scaffold on which they depend can have devastating consequences on neuronal function.

6.6 Neuroinflammation

Inflammation of nerve tissue can be caused by injury, infection, stroke, toxic metabolites or autoimmunity. Inflammatory processes can reduce available metabolites, increase oxidative stress, disrupt intercellular contacts and tissue architecture, synaptic pruning, destruction of cell extensions and phagocytosis, such as myelin, and even lead directly to apoptosis glial cells and neurons. Acute inflammation can severely impair neurological function, although it is usually reversible. Recurrent or chronic inflammation often leads to progressive neurodegeneration.

6.7 Demyelination

Loss of myelin, which protects nerves due to trauma, toxins, metabolic changes, or inflammation, exposes axons to environmental damage, leading to short-term neuronal dysfunction and loss fast welcome signage. Demyelination dramatically increases energy requirements and alters cell signaling. Chronic demyelination leads directly to Neurodegeneration.

6.8 Glial dysfunction

Neurons depend primarily on glial support. Glial cells such as astrocytes and oligodendrocytes (or Schwann cells in PNS) provide important cell signaling and metabolic support to neurons and axons. Glial cells also remove cellular waste products, such as toxic metabolites and ROS, from neurons. Loss of glial cells or glial function and acquired glial aging can accelerate or completely lead to neurodegeneration.

6.9 Induction of cell death

The induction of cell death, in particular the triggering of apoptosis, is one of the most important factors leading to neuronal death in neurodegenerative diseases. Apoptosis can occur externally, mainly during inflammation of immune cells. However, it can also be caused by extreme damage to mitochondria, overload of the autophagy pathway, or damage to cellular structures caused by protein aggregation.

7. Diagnosis

This is a very important concept as many efforts are being made to develop biomarkers for diagnosing these diseases and monitoring disease progression in clinical trials. A multidisciplinary approach is commonly used to improve the quality of life of people with neurodegenerative disorders. Research continues to find new, much-needed treatments for neurodegenerative diseases. One of the most exciting treatments is using stem cells to replace dead neurons. There are too many smart people.

7.1 The burden of neurological disease

The burden is already large and growing. It is currently estimated that up to one billion people worldwide suffer from neurological disorders and their consequences. These disorders occur at all ages and locations. The number of people with neurological disorders is expected to increase significantly in the coming years. The number of people with dementia (already in the tens of millions) is predicted to double every 20 years. Physical therapy guidelines have been published that systematically review the best available evidence for these conditions. "Most patients are referred for this treatment by a physical therapist. See Individual Physical Therapy Certification."

8. Future scope and prospects

Despite all advances in genetics, bimolecular science, and pharmacology, effective treatment of noncommunicable diseases requires identification and risk reduction, complete cure, or at least long-term relief of symptoms. Although some strategies have been developed or may be applied to lower animals, clinical trials and use in humans have not yet been completed. One of the new treatment strategies include: Inhibition of disease-related protein deposition. Protein misfolding occurs due to gene mutations, oxidative stress, aging, and changes in cell temperature and pH.

9. Conclusion

Neurodegenerative disorder is the progressive death of neurons under the influence of environmental, biochemical, genetic, and epigenetic factors. Free radical generation and deterioration of the antioxidant system due to oxidoreductase activity have been shown to cause aggregation of misfolded proteins in the central nervous system, leading to mitochondrial dysfunction and neuroinflammation, ultimately leading to NCD. Freud developed a series of psychotic techniques that involved the use of cognitive development strategies such as free association, daydreaming, and the like. Creative imagination governs the formation of dreams according to principles, so the interpretation of literature can govern dreams as follows. Harmonize the process. The aesthetic pleasure that the artist gives us lies in the happy pleasure that relieves the tension of the mind. It allows us to enjoy our own dreams without shame. This brings us to the threshold of a new, interesting and somewhat complex action by the end of the discussion.

Further reading

Aboulafia-Brakha, T., Suchecki, D., Gouveia-Paulino, F., Nitrini, R., & Ptak, R. (2014). Cognitive-behavioural group therapy improves a psychophysiological marker of stress in caregivers of patients with Alzheimer's disease. *Aging and Mental Health, 18*(6), 801–808.

Alexander, N., Alexander, D. C., Barkhof, F., & Denaxas, S. (2020). Using unsupervised learning to identify clinical subtypes of Alzheimer's disease in electronic health records. *Studies in Health Technology and Informatics, 270*, 499–503.

Bovolenta, T. M., de Azevedo Silva, S. M. C., Arb Saba, R., Borges, V., Ferraz, H. B., & Felicio, A. C. (2017). Systematic review and critical analysis of cost studies associated with Parkinson's disease. *Parkinson's Disease.* , Article 3410946. https://doi.org/10.1155/2017/3410946

Dujardin, S., Commins, C., Lathuiliere, A., Beerepoot, P., Fernandes, A. R., Kamath, T. V., De Los Santos, M. B., Klickstein, N., Corjuc, D. L., Corjuc, B. T., et al. (2020). Tau molecular diversity contributes to clinical heterogeneity in Alzheimer's disease. *Nature Medicine, 26*, 1256–1263. https://doi.org/10.1038/s41591-020-0938-9

Garg, M. (2021). *Quantifying the suicidal tendency on social media: A survey.* arXiv. preprint arXiv:2110.03663.

Goodfellow, I., Bengio, Y., & Courville, A. (2016). *Deep learning.* MIT Press.

Hinton, G. E., & Salakhutdinov, R. R. (2006). Reducing the dimensionality of data with neural networks. *Science, 313*(5786), 504–507. https://doi.org/10.1126/science.1127647

Katsnelson, A., De Strooper, B., & Zoghbi, H. Y. (2016). Neurodegeneration: From cellular concepts to clinical applications. *Science Translational Medicine, 8*(364), 364ps18. https://doi.org/10.1126/scitranslmed.aal2074

Kaur, S., Bhardwaj, R., Jain, A., Garg, M., & Saxena, C. (December 2022). Causal categorization of mental health posts using transformers. In *Proceedings of the 14th annual meeting of the forum for information retrieval evaluation* (pp. 43–46).

Mbuba, C. K., Ngugi, A. K., Newton, C. R., & Carter, J. A. (2008). The epilepsy treatment gap in developing countries: A systematic review of the magnitude, causes, and intervention strategies. *Epilepsia, 49*(9), 1491–1503. https://doi.org/10.1111/j.1528-1167.2008.01693.x

Sköldunger, A., Johnell, K., Winblad, B., & Wimo, A. (2013). Mortality and treatment costs have a great impact on the cost-effectiveness of disease modifying treatment in Alzheimer's disease: A simulation study. *Current Alzheimer Research, 10*(2), 207–216.

Valliani, A. A.-A., Ranti, D., & Oermann, E. K. (2019). Deep learning and neurology: A systematic review. *Neurology and Therapy, 8*(2), 351–365. https://doi.org/10.1007/s40120-019-00153-8

Woods, B., Aguirre, E., Spector, A. E., & Orrell, M. (2012). Cognitive stimulation to improve cognitive functioning in people with dementia. *Cochrane Database of Systematic Reviews, 2*, CD005562. https://doi.org/10.1002/14651858.CD005562.pub2

Index

'*Note:* Page numbers followed by "f" indicate figures and "t" indicates tables.'

D

E

P

Q

R

S

Printed in the United States
by Baker & Taylor Publisher Services